# See for yourself

# See for yourself

A guide to the use of video by the Christian community

Main contributors:
Anton Baumohl
Jim Eglinton
David Hamilton
Ken Jackson (editor)
Alan Matthews
Stephen Maxted
Leslie Timmins

Jay books

**Jay Books**
Woodside, Hadlow Park, Hadlow, Tonbridge, Kent TN11 0HZ, England

First published 1985


**British Library Cataloguing in Publication Data**

See for yourself: a guide to the use of video
by the Christian community
1. Video tape recorders and recording
2. Great Britain – Church history – 1945 –
I. Jackson, K. G.
274.1 BR743.2

ISBN 0-9510086-0-9

Printed by The Whitefriars Press Ltd, Tonbridge, Kent

# Preface

To many in the church video is still a novelty – understandably so because the medium has been available for only a very few years. A consequence of this is that there is no great body of experience to draw on.

As people have experimented and acquired some know-how they have sought to pass it on to others. This has resulted in numbers of video seminars based on individual experience. The need has been apparent for a publication that pulled together the lessons learned in different circumstances and in different applications – for it is clear that video has lots of uses for lots of people but that different approaches are often needed.

So far, because of the youth of the medium, there are few recognised experts. The six contributors to this book have been selected as prime advocates of the use of video within particular areas, but they would be the first to agree that even their knowledge is incomplete.

Each has been asked to provide the benefit of his experience and inevitably there is some overlap in this. To edit out all repetition would risk eliminating some useful angles and insights, and so each author provides advice on his speciality in his own way and from his own viewpoint.

There are uses of video that have yet to be fully explored, and in these areas we have to draw the best sketch map that is possible as a guide to those who will be following. On the other hand there are applications where video is *not* the best medium to use, and these are clearly indicated.

I believe that you will find the following pages helpful and that you will be encouraged to take up the challenge and opportunities of using video to encourage God's people, strengthen the church and spread the gospel in a new and exciting way.


## Acknowledgements

In addition to the contributors and organisations named on the opposite page, thanks are due to many others who have directly or indirectly provided ideas, facts, case histories or general information to make this book complete. They include Geoff Bone (Impact Ministries), Rev. Steve Chalke, Nigel Cooke (International Films), Trevor Jackson (Gospel Vision), Trevor Payne (Christian Video Experience), Rev. Fergus Robertson and Jean Wilson (Mission England) as well as Hilary Norman and Frank Mancktelow who designed and photographed the cover.

KJ

# Contents


1 What is video? 9
Ken Jackson

2 Video for housegroups 22
Alan Matthews (Bagster Video)

3 The Christian life 33
Leslie Timmins (CTVC)

4 Training for service 47
David Hamilton (Church of Scotland Education)

5 Working with children and teenagers 62
Anton Baumohl (Scripture Union)

6 Video in evangelism 78
Stephen Maxted (Spring Harvest)

7 A video ministry 94
Jim Eglinton (Highland Christian Video Fellowship)

8 Religious entertainment! 103
Ken Jackson

Index 115

# 1

# What is video?

## Ken Jackson

My milkman wouldn't think of going to church. But he was delighted to be offered a video of *The Hiding Place* to watch in the evening with his family. When he returned it, he was equally pleased to take *Joni* for the next evening. Like some 30 per cent of the population he has a video recorder and is prepared to watch, in his own home, films that he would not dream of going to the cinema for.

The capacity for video consumption is enormous, particularly among the impressionable young. When a London City Missioner promised an evening of video, he was heartbroken at the crowd of children who turned up clamouring for 'nasties' – films of violence and horror. It is an awesome responsibility that we have to offer these children something better and to demonstrate that 'good' programmes can also be entertaining and well made.

Enlightenment of our neighbours is not the only use for video. In the following chapters we attempt to show the various other ways in which it can be used and how it can be used most effectively.

But first –

## What is video?

In the popular sense that we are using it here, it is 'canned' television. When broadcast television first arrived it brought the facility into the home of seeing famous people at work. It enabled the housebound to attend concerts, lectures, church services and outdoor events. As the medium developed, we became able to see what was happening in other countries, with first hand reports from trouble spots, mission fields and disaster areas. It provided the wherewithal for a better informed audience. One major limitation however was that of timing. Unless we could be in front of the 'box' when the programme was broadcast, it was lost to us for ever (except for possible repeat showings at the discretion of the broadcast authorities).

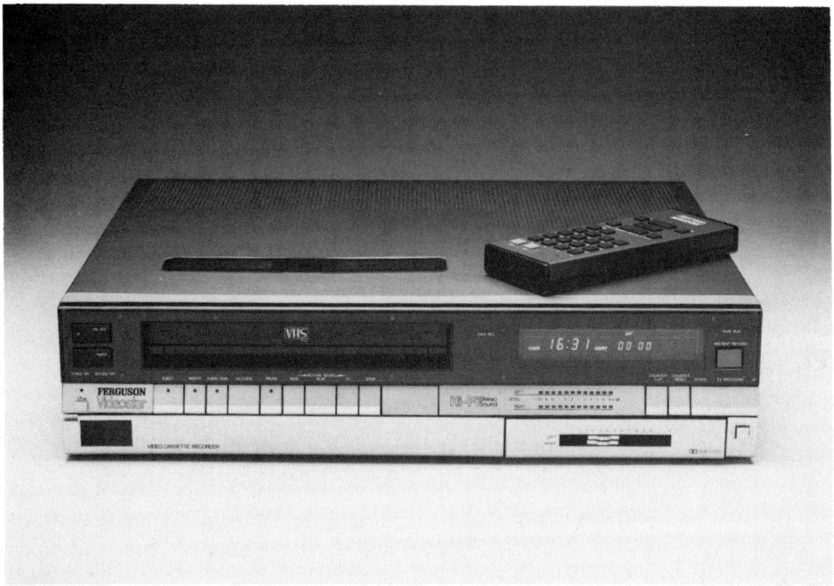


*This VHS video cassette recorder is front loading. On the right of the slot for the cassette is an information display including a clock and counter. Timer controlled recording information appears here. The row of switches below the slot provide for tape ejection, recording, audio dubbing, picture freeze, fast forward and rewind while viewing the picture, and straightforward replay. The Videostar model shown provides stereo sound from the increasing number of commercial videos that incorporate it. A flap beneath the row of switches hides controls that require only occasional adjustment (Ferguson)*

The advent of the video recorder changed that. First and foremost it has been used as a time-shift device, to enable programmes to be stored and viewed when convenient. Not only can we see programmes that are broadcast in our absence, we can also see those that have coincided with others on another channel. *Songs of Praise* on BBC 1 on Sunday evening usually coincides with an ITV programme of similar interest. With a video recorder, you can watch both!

Having acquired a recorder, however, there is much more that can be done with it. Pre-recorded tapes that can be borrowed or bought enable you to see all sorts of programmes that are never likely to be broadcast because of their minority interest or their unsuitability for the mass medium. Video libraries have in the past attracted unfavourable attention because of the very diverse nature of the material that they offer on video. Some regulation is now being brought to this by certification schemes.

In this field of enormous challenge, Christian organisations have not been idle. Although the medium is not very old, a number of highly professional dedicated Christian producers have emerged who are making video programmes that compare in quality with the best secular material. Many others are trying their hand at providing Christian programmes for various purposes. The *Guide to Christian Video* contains details of well over a thousand programmes from a hundred or so different organisations.

## Why use video?

But, you may be saying, we already have a film projector and a slide projector. A lot of good material is available on film and film strip. Why spend more?

This is a valid question. There are situations where you want to present visual material on a large screen to a big audience. The picture quality that you will get from 16 mm film or 35 mm transparencies will be better than that from a similarly enlarged video picture. The nature of the process means that there is a line structure across a video picture. The image is constructed from 625 lines (525 in America or Japan) and these become more obvious as the picture is enlarged.

Also, as you may just have gathered, there are various standards for video pictures in different countries (more on this later). So you may encounter difficulties if you want to show a programme originating abroad. Ciné film on the other hand will cross frontiers without this difficulty.

However, there is now material on video that is not available on film, and in many situations (such as a housegroup) video is much more appropriate and requires less setting up. Also, it is very unlikely that you will have to spend any money at all on equipment to show a video (unless you are planning a large-screen presentation to a large audience). There is probably at least one member of your group who has a video recorder and who will be willing to let you use it. You may have a hire charge for the video cassette (as you would for a film), but quite a few organisations offer free loan of their own productions.

## When should I use video?

We have mentioned one situation where a film might be preferable. Although video is very versatile, it has its proper place, and there are many occasions when it would not be appropriate.

Its ease of use may tempt us to use it as a soft option, to evade our responsibilities. In the presentation of the claims of Christ there can be no substitute for an effective speaker making a personal challenge

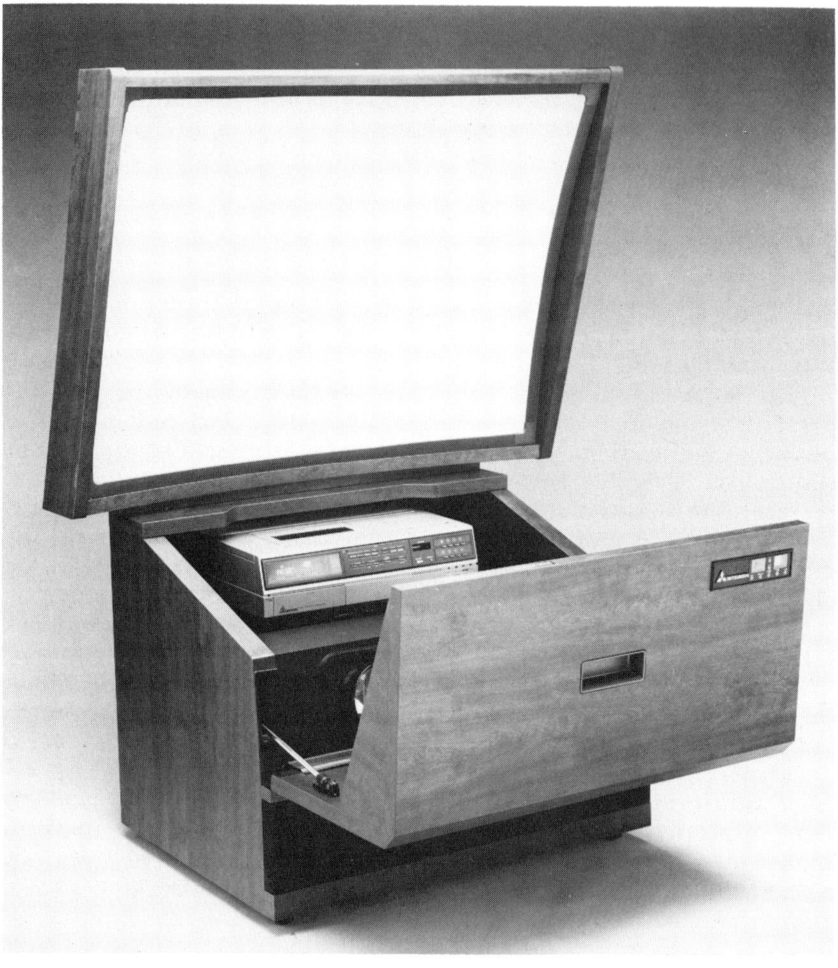

*One-piece projection television. The images from three separate picture tubes (one for each colour) are reflected off a mirror in the lid, which is shown hinged forward, to converge on the 50 inch screen (Mitsubishi)*

demanding an immediate response. On the doorstep, witnessing to our neighbours, it is our own lives, words and actions that are judged. However, a carefully selected video programme can help to prepare the way for a personal approach.

In training applications, a video alone could produce a passive response. It needs to be supplemented by work material, discussion, etc.

For a house meeting or discussion group, a speaker may be available on an audio tape. Is there then any advantage in using a video version? After all, it is the words that we are interested in. Experience shows that an audience is able to concentrate more on a visual presentation. While listening to an audio tape we let our eyes wander around the room, seeking or avoiding other peoples' eyes, noticing items of furniture. A video gives a focus for attention within that familiar, friendly television set!

## How does it work?

A video recorder is designed to receive a broadcast signal from an aerial and to record it and/or pass it on to a television receiver for viewing. It records the broadcast signal on to a belt of magnetic tape that is housed in a cassette (just like an audio cassette but rather larger).

Unlike an audio tape recorder, there are several different systems which are not compatible with one another.

Firstly, there are three main forms of television broadcasting and reception around the world. In Britain, much of Europe, Australia, New Zealand, South Africa, most of South America and China, the system is PAL (Phase Alternate Line). North America has a system known as NTSC (after the organising body). France, Russia and part of Africa use SECAM.

The recorder used will always need to conform to the broadcast system of the country in which it is located, and programmes recorded on a machine of one system will not be capable of being replayed on a machine of another system. Even within the PAL group of countries there are variations in detail so that a programme recorded off-air in Britain cannot be played on a German machine (and vice-versa) without loss of sound. If programmes from one system are to be used on another they must first pass through a conversion process to produce a compatible tape.

If you plan to show many programmes that you import from another area, you can obtain (at some cost) a multi-standard player which will accept cassettes of the three different systems.

So we know that we are restricted to programmes recorded on our own system. This is not the end of the matter. Manufacturers of domestic video recorders have produced three competing, and incompatible, processes. The most widely used, and the one in which virtually all tapes can be obtained, is known as VHS. The second, and its devotees say the best, is Betamax. The third, which possibly is the best but which has suffered from being the last to be developed, is Video 2000.

Each of these systems has its advantages and its loyal supporters. However market forces ensure that programmes are most widely available for that system which is present in greatest numbers. VHS has 80 per cent of the market and this ensures that video libraries stock VHS type tapes.

Any of these three domestic systems is adequate for playing a good quality video tape and converting it into an acceptable picture on the screen of your television set. It is even possible, by the use of two video recorders connected 'back to back', to copy one of these tapes and still have an acceptable picture from the copy. (But please note that a copyright programme should not be copied without

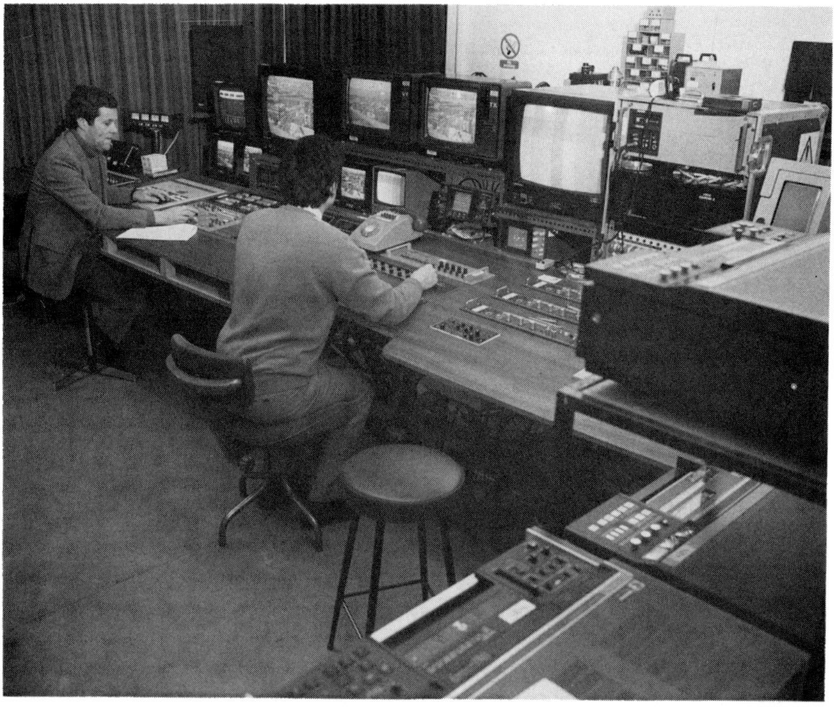

*The editing desk of a video production unit. Ian Gall and John Williams are preparing a master tape (Church of Scotland)*

permission. There is more about this at the end of the chapter.) Every time a tape is copied, however, there is some reduction in quality. Professional producers therefore have to start with a very high quality tape so that the result after several necessary editing or copying stages (often called 'generations') is still acceptable. The professional recording format used by most Christian video producers is U-matic. It employs a ¾ inch wide tape rather than the ½ inch tape that is in the VHS, Betamax or V2000 cassettes. We mention it here because a few libraries do offer U-matic format tapes and some users do prefer to have the professional standard U-matic recorders.

While we are considering tape sizes, we might just mention also that ¼ inch (or 8 mm) tape systems are available for portable video cameras but there is no advantage in using these for pre-recorded programmes.

The video cassettes used by the three domestic ½ inch tape systems are all approximately (but not exactly) the same size. They consist of a plastics case containing the spools and a guard to protect the tape from accidental handling. A blank tape cassette comes with a tab that may be broken off after recording to prevent accidental subsequent erasure. A missing tab on a recorded cassette may be substituted with a strip of adhesive tape if it is required to wipe off the recording. The magnetic tape itself is vulnerable to finger marks, dust and stray magnetic fields; the case is liable to warp if it gets unduly warm or is left lying on its face. Some hints for their care are listed.

*Care of video cassettes*

Store cassettes in an even, moderate temperature in a dust free enclosure.

When not in use, cassettes should be kept in the sleeve provided (or in a library case) with the tape edge farthest from the opening.

Rack cassettes in a vertical position to minimise the risk of them becoming warped or distorted.

Warping can also result from heat. Do not put cassettes on a warm surface such as a tv set.

Avoid touching the tape with fingers. Grease on the oxide surface can harm the video heads.

Loudspeakers, headphones and tv sets include magnetic sources which can cause partial erasure of a tape.

Use the *pause* control sparingly as it increases the wear on a short section of the tape.

Take care when *fast winding* cassettes. Operate the control firmly to avoid loops of tape becoming slack and creased.

To reduce handling, label every cassette or sleeve. Re-usable labels are available.

Once the recorder is purchased, we are committed to a particular format. The principles of operation of a video recorder are similar to those of an audio tape recorder (but more complicated!). The programme is recorded onto a magnetic tape when it passes over a recording head. An adjacent playback head reads the recording back off the tape and presents it to the television screen to create the picture.

The recorder contains a tuner unit and will record a television programme as it is being broadcast, whether it is being viewed on the television screen or not. So you can record a programme without turning your tv set on, or (by using the built-in timer) while you are absent, or while you are watching a programme on a different channel.

Your television receiver probably has a channel reserved for video use, and the operating manual (or the service engineer) will explain how to tune it. The incoming lead from the aerial is plugged into the video recorder and the output from the recorder goes into the aerial input socket of the tv set. The equipment is then ready to record broadcast programmes or replay pre-recorded cassettes.

There are many more facilities available on a video recorder than on an audio tape recorder. On most machines you have buttons that enable you to *search* backwards or forwards to find a point you want in a programme. Most valuable when a programme is being used in a teaching situation is the *pause* button, which stops the programme and holds the picture (more or less) still on the screen for as long as required. Do not over-use the pause control however because it does cause excessive wear of the tape.

*Fast forward* or *rewind* enable you to move more quickly to a part of the tape that you want. The picture is not available on the screen while under these controls, but most recorders have a tape counter. If you want to identify a point in a programme for future reference, set the tape counter at zero at the start of the programme and note the reading when the point is reached. This will enable you to find it again quickly when it is wanted.

In a subsequent chapter, David Hamilton suggests the use of extracts from several programmes during the course of a training session. To do this requires the accurate location of the required extract, and the tape counter will help with this. If a tape contains a number of items that you may want to refer to, compile an index on the cassette case, listing the items and their counter readings.

Tapes are available in lengths that will run for periods from 30 minutes up to four hours. Before recording a programme ensure that you have a sufficiently long tape in the machine. Before using a tape

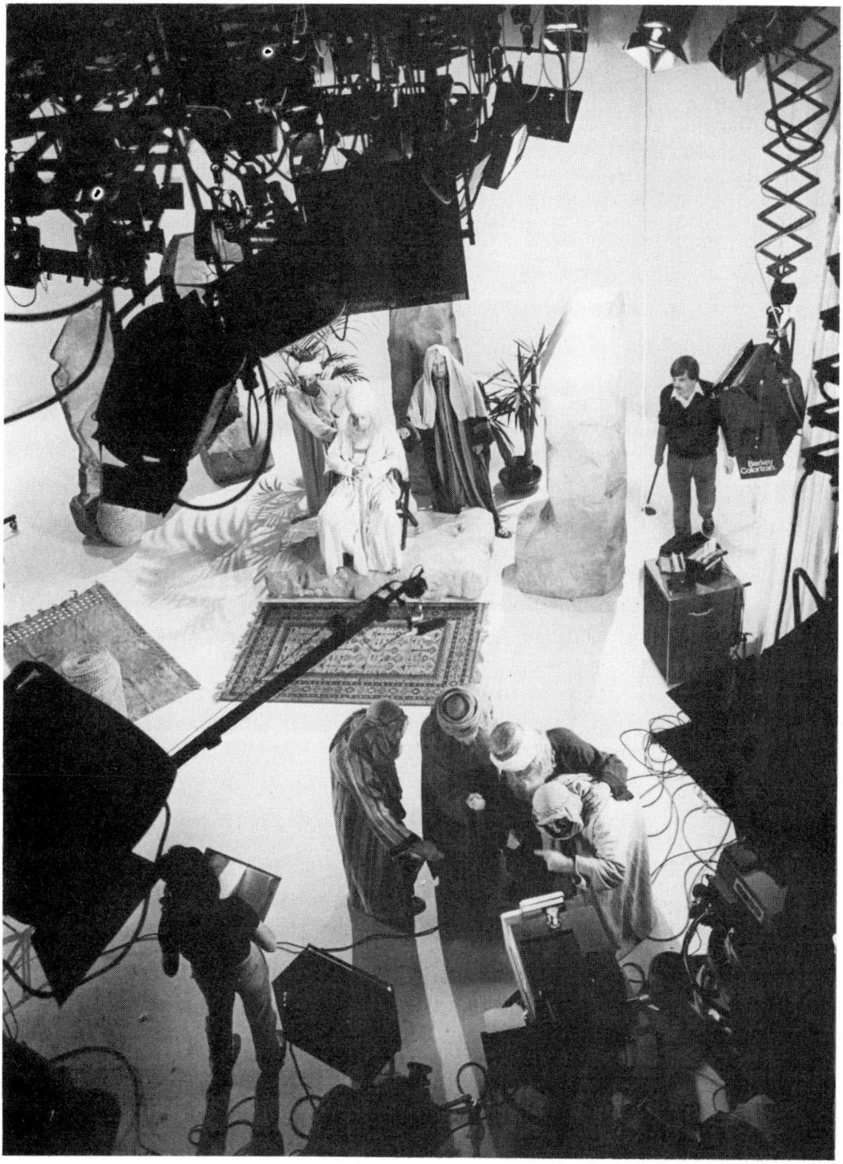

*The scene in the studio during the making of a biblical video (CTVC)*

in a meeting, check its running time. Nothing can be more disconcerting, or counter-productive, than to have to cut a speaker off before he reaches his climax so that the audience can catch the last bus or meet the children from school. These can seem simple and obvious points, but they are surprisingly often overlooked!

One other control should be mentioned. The knob for setting the *auto-tracking* is sometimes concealed behind a panel with other less frequently used controls. Sometimes when you replay a video that was not recorded on your machine there are broken black and white lines across the top or bottom of the picture. These are the result of a tracking difference with the recording machine and may be eliminated by adjusting the auto-tracking control.

### *The screen*

For viewing at home, an ordinary television receiver of up to 26 inches will probably be adequate. This would also serve in a small hall with an audience of up to 30 people.

As your audience gets larger you have either to increase the number of television screens or to get projection equipment. Several television sets positioned down each side of the hall is less than satisfactory. The video signal can be divided between the sets by means of an aerial splitter which can be obtained from the local tv dealer to enable one aerial to feed two television receivers. However the floor becomes covered with a network of vulnerable cables and the centre of attention is moved away from the centre of the room, where the leader or presenter may be, and turns eyes out to the periphery. It is worse for those sitting in the middle who may be encouraged to look first to the left and then to the right to compare the pictures.

Far less distraction, or neck craning, is caused by a tv projector. This is available with a screen size from 40 inches up to 36 feet. (All screen dimensions are given as measurements of a diagonal.) For 75 people, a 40 inch screen should suffice; for 120, a 50 inch screen; for 250, a 78 inch screen.

There are three basic types of system: (a) floor models (or two-piece systems) in which the screen is placed at a distance from the projector so that the picture is beamed directly onto the screen rather like a film projector; (b) rear projectors (one-piece) which have a mirror to compress the lens-screen distance; and (c) reflex projectors with two reflecting surfaces.

*Two-piece systems.* The screen of a two-piece direct projector may be mounted on a stand or attached to a wall. Avoid a position where direct light may fall on its surface, and particularly steer clear of fluorescent light if possible. Outside an angle of about 30° on either side of the centre line of the screen the image brightness and the clarity of viewing will fall off, so in a rectangular room try to place the screen in the centre of a shorter wall or else in a corner. But do not put it in front of a window so that the projector lens is shining into the sunlight.

The projector unit normally sits on the floor at a distance from the screen dependent on the screen size. For a 60 inch screen, about 6½ feet is needed; for a 100 inch screen, it will be over 11½ feet. One manufacturer supplies a 100 inch screen with a ceiling mounted projector.

There are two types of screen. The cheapest are flat. If these have a pearlux or beaded surface they will provide a reasonably bright image with less fall-off at extreme viewing angles. Parabolic screens are highly reflective. They can give a bright image suitable for viewing in a high ambient light. They are of course more expensive and are limited in size to 100 inches.

*One-piece systems.* These are self-contained and take up less space. In fact some are used for domestic viewing. The screen sizes do not exceed 50 inches but the picture quality is good and reasonably tolerant of the lighting conditions. Because of its unitary construction, the screen cannot be moved and so it is necessary to ensure that the position of the set is absolutely right for all the audience before the meeting begins.

A reflex projector usually contains a hinged mirror and rear projection screen. One model has a screen that automatically retracts into its cabinet when the receiver is turned off, so making it into an attractive item of domestic furniture. The maximum available screen is 60 inches.

### *The projector*

You may have wondered at the references above to projector lenses. To obtain a bright picture on a screen some distance away, the television picture is separated into its three primary colours and each is projected through a separate tube. The projector therefore consists of three tubes – red, green and blue. To get a sharp picture each must be accurately focused on the screen and also the three colour images must overlay each other precisely to give sharp definition without any colour fringes.

For the smaller units, this focusing and colour registration is performed simply and almost automatically. The 10 foot and larger screens require variable-focus projectors.

The costs of purchase of any of these equipments will run into several thousand pounds, but they can be hired by the day, week or month for prices of £100 upwards. Some will need an operator, or a technician to install. Time should be allowed for this before the audience starts to arrive.

## Software

Having chosen the right equipment (or 'hardware') for the job, we need to give careful thought to the 'software' or programmes that we are going to show on it.

Perhaps because of the newness of video, some people are surprisingly uncritical about what they see. I have been in a Christian bookshop when a customer has come in and asked for 'a video' to show at a meeting. Would you expect to go to that shop and ask for 'a book' to read? You would of course have given some prior thought to the nature of the book you wanted, such as to assist your bible study, to learn more about a particular mission field, or just good reading for relaxation. You will know how deeply you want to go into its topic. You may even know the author you want.

There is just as much variety available in Christian video. Do you want your programme to appeal to a non-Christian audience? If so, how old is the audience and what are its likely interests? Would Cliff Richard turn them off? Would a dramatic story like *The Cross and the Switchblade* be more appropriate? (This latter tape has been widely and successfully used with teenagers, but its sixties style is now beginning to handicap it.) Do you want to challenge your viewers, or teach them? What do you want to teach them? Will the meeting run for half an hour or two hours? Should there be a break in the middle?

To start your search for the right programmes, they are all classified and listed (with running times) in the *Guide to Christian Video.* This will also direct you to the producers or distributors, who will be glad to let you have full details of the contents of their videos, and the terms on which they are supplied.

Most will be available for hire, and many Christian video libraries are springing up, although most have only a small selection from the many tapes available.

## Copyright

Finally, a warning about copyright. Copyright indicates ownership, and it exists in every pre-recorded video programme. To make a copy of a programme without the prior permission of the copyright holder is to steal his work.

It is tempting sometimes to argue that the work of the Lord requires the possession of a particular video. In that case especially, it should be acquired honestly.

Copyright also exists in broadcast programmes. The sale of video recorders indicates a general acceptance that these programmes may be recorded. However, although subsequent viewing at home by the family may be tolerated, there could be more serious repercussions if the programme were offered for sale (or hire) or even shown at a public performance.

Most video libraries have their own set of rules or conditions under which their programmes may be used. It is good stewardship to make sure that you understand and conform to them.

2

# Video for housegroups

Alan Matthews (Bagster Video)

Using a video in a housegroup meeting will no doubt appear to some as the very antithesis of that type of gathering. However as you are reading this book I am working on the assumption that the idea appeals to *you* and that you are keen to utilise every form of *teaching tool* that is available to *reinforce your message.*

Now there were a lot of italics in that first paragraph. This is deliberate because I want you to see that to make video work successfully in a housegroup situation it is not enough just to switch on the recorder, relax and watch the programme!

The video cassette is a tool for the Christian to use (probably the finest since the blackboard was invented) and its use needs to be planned so that all can benefit by not only assimilating the information contained in the programme but also appropriating the message; that is by taking it in and applying it to our lives. Only then will we start to realise its full potential.

If you are a housegroup leader, don't ever look upon video as a substitute for the role that the Holy Spirit has called you into. Always plan your programmes so that full and adequate follow up can be accomplished by your group. Look upon the video as a starter, then when you switch it off, tuck into the main course!

Of course, there are times when you shouldn't use video at all. Be sensible and sensitive in this area. It is all too easy to 'cop out' and let the machine do the work which deep down you know you should be applying yourself to. To be able to reach out to your neighbours with videos like *Joni* creates fantastic opportunities. It is not enough to enquire of them if they 'enjoyed' the film. Engage them in conversation and draw from them their reaction to one girl's faith amid hardship. Apply it to themselves and then lovingly tell them of a saviour who cares for them as much as he cares for Joni.

Let us now look at some of the ways that video can be used in a housegroup situation. The commonest format is VHS which stands for Video Home System, so you can see that it is ideally suited for housegroups!

## Teaching

Teaching is one of the major functions of a housegroup. It is not to be confused with Education, which is dealt with elsewhere, but is allied to bible study as the group searches out the truths and lessons of God's message. There are many fine teaching programmes available and I am confident that there will be lots more soon as more and more Christian production houses start to bring their products before the public's eye.

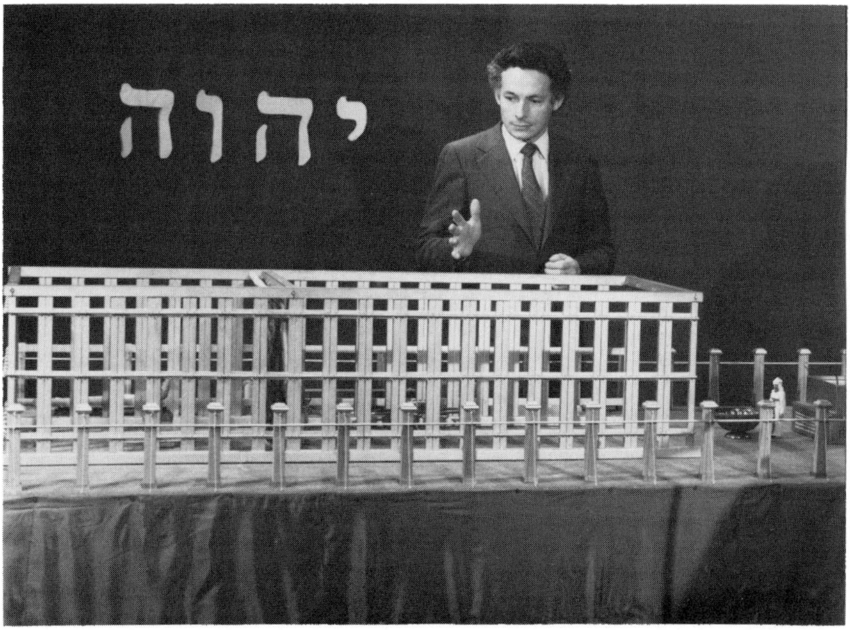


*Christopher Smith stands behind the model to explain* The Tabernacle *(International Correspondence Institute)*

Most teaching programmes tend to be in the form of a series of tapes. A good example is *Genesis* by David Pawson. This is a five tape set with two three-quarter hour programmes on each cassette yet covers only the first three chapters of the Bible! With in-depth teaching of this sort, one quickly begins to realise that the pause button on the video recorder is an invaluable addition.

Using your home for evangelism and/or teaching is rewarding in itself, allowing you to get to know each individual on a personal basis and to help one another towards greater maturity in Christ. Having access to some of the great men of faith that God has raised up in our age is an added bonus from which we can all benefit if we use the tapes intelligently. Here are some guidelines for the leader:

1. Make sure that you have previewed the programme. That way you will not be caught out by the contents (or the running time).

2. Sometimes teaching programmes have study notes accompanying them, but if they do not, try to think of some pertinent points to bring out, so that progression into discussion isn't too difficult. Be sensitive to the needs of your group; you know them best; adapt the content of the message to their requirements.

3. Be sure to allow time for discussion. Too often in housegroups an evening is wasted in peripheral activities. Remember that you have all gathered to pray and study God's word. Make certain that you allow time to do just that. Catching up on local news can wait until later – much later!

4. If you are studying a particular book or section of scripture over a number of evenings, encourage the group to do their part by preparing prayerfully at home for the study. Where possible all should try to come to the session having at least read the passage to be expounded. It is rewarding to see a housegroup apply itself to biblical education through active participation rather than by passive presence.

5. Be prepared to switch off the tape before the end. Remember that it is only a tool. Allow the Holy Spirit to complete this work and be sensitive to him.

6. Finally, try to encourage your group to *appropriate* what they have learned. They should apply the principles they have been taught and reach out into their community with a living and dynamic faith centred around the Lord Jesus Christ.

Here are some programmes that I can recommend for teaching sessions. The list is not comprehensive and there are many more that are worth viewing, but it will give you some idea on commencing a video bible study hour.

The series on *Genesis* has already been mentioned. David Pawson provides ten 45 minute talks on the contents of the first three chapters of the Bible ranging from 'The Real Conflict' and 'The Beginning' through to 'Enter the Devil' and 'Paradise Lost'. He has been described as one of Britain's finest biblical communicators.

*The Tabernacle* is the title of a tape from Videout by Billy Strachan of Capernwray Bible College. There is another programme with the same title in the Living Word series by ICI Bible Study. This latter series of six half-hour programmes is intended for serious students and is narrated by Christopher Smith of Elim Bible College.

*A Mind Behind it All* is a study on John by the late Professor Blaiklock. It is a series of six quarter-hour programmes filmed during the last four months of the professor's life. He draws on his vast classical and historical knowledge to reaffirm his lifelong belief that "there must be a God, that Jesus Christ must be his last word to man, and that the new testament is true and is the Word of God". The tapes are accompanied by a leader's guide.

*The Measure of . . .* series by Dr Gene Getz are all based on the letters of Paul. There are five self-contained half-hour programmes on The Measure of A Man, A Woman, A Marriage, A Family, and A Church. Dr Getz reveals the essence of Paul's teaching on each of the subjects, raising questions for discussion and prayer. Scripture references are displayed on the screen.

George Verwer, the founder of Operation Mobilisation, is the speaker in *Literature Evangelism* and emphasises the value and importance of Christian literature. He claims that literature has played a major part in every spiritual revolution.

## Missionary meetings

Missionaries for me are a bone of contention! It is not that I don't like them, it is in fact quite the opposite. I appreciate them and look forward to meeting them and hearing of what God is doing in the situation in which he has placed them. It is just that I cannot help thinking how much they ought to be enjoying a rest (isn't that what furlough is for?) and not tiring themselves out doing the 'rounds' of their prayer supporters.

Now that is a real problem, because in order to pray effectively for a mission or an individual you need to have first-hand reliable information. This is almost another Catch 22 situation.

SIM, Wycliffe and other missionary organisations have begun to realise the value of putting their message onto video and sending out *that* to potential supporters and prayer partners *as well as* real-life missionaries. I believe that is a real step forward in not only giving an up-to-date visual presentation but also actively reducing the work load on already over-committed workers.

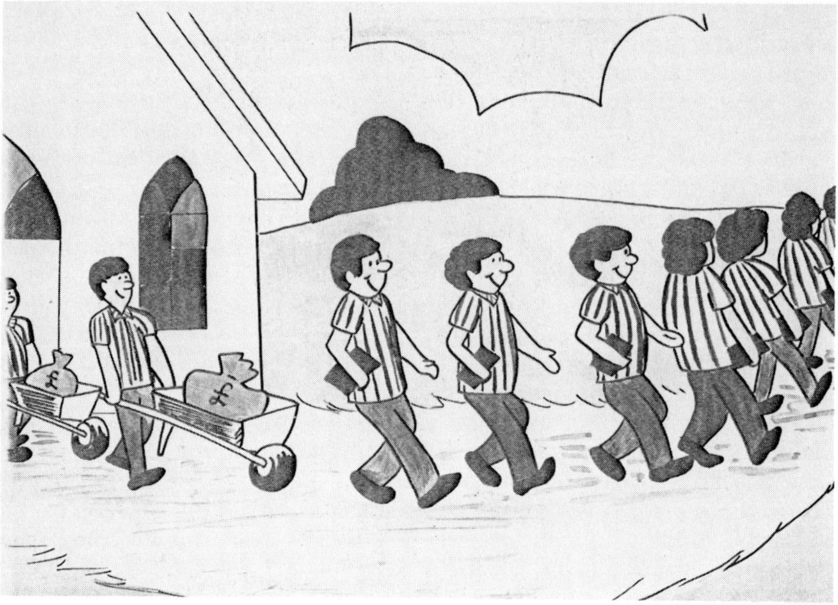

*Missionaries set out with financial support from the church. A scene from* Who Sends Them *(SIM)*

SIM video producer Derek Frost has related to me how his church recently sent a worker to Africa. A short while later SIM sent Derek to that area to help the locals use video and also to get footage for future programmes. In a brief interview, filmed in the middle of the night in a hospital ward whilst on duty, the young mission worker was able to relate to her church just how she had felt over her first few weeks, going through all sorts of emotions while struggling with a new language, weather conditions and the general culture shock of her new environment.

Derek told me that the impact of that short video on the church was tremendous, and the interest generated in *their* worker was considerable. This was just one more aspect of using video tape intelligently.

There are not very many educational missionary videos available yet but those that are are well worth watching. Try to catch the following if you can. God may use one of them to call another worker.

*Faces of India* is a documentary that portrays the faiths of India – Muslim, Hindu and Christianity – in an objective manner in the context of everyday life. It is a report on the work of the OMS and was shot entirely on location by the Christian Television Association.

Tear Fund have produced a three-part cassette which includes *Cliff*

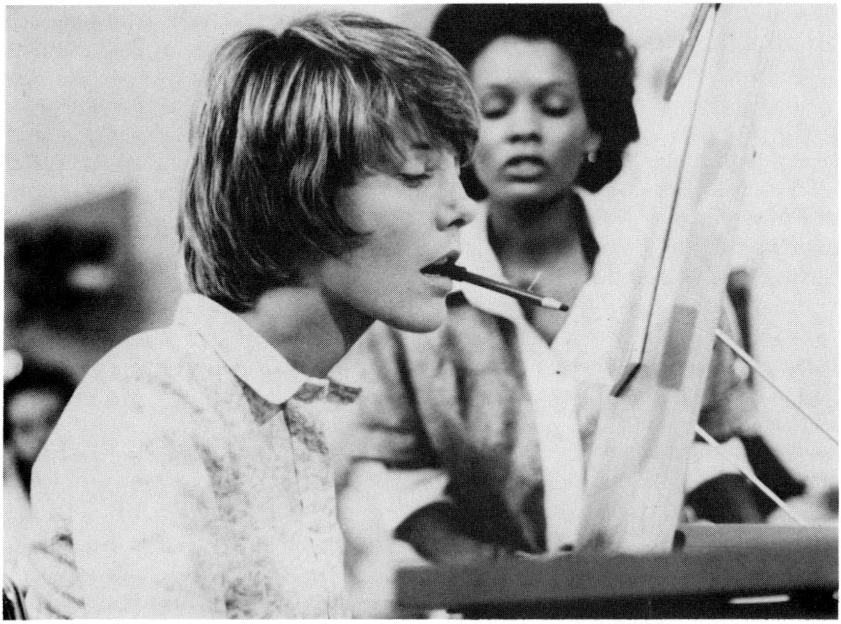

*It was through the encouragement of one of Joni Earackson's physical therapists (played here by Cloyce Morrow) that she began writing and painting with a brush held between her teeth. Joni portrays herself in* Joni *(International Films)*

*in Kenya* (Cliff Richard, that is) and his fellow musician Garth Hewitt in *India.*

A challenging film on the relationship between the local church, the missionary and the missionary society is the cartoon video from SIM entitled *Who Sends Them?* This ends with a selection of penetrating questions on the role of the local church in providing missionary support.

Other missionary societies are preparing videos of their work. One from SAMS is listed at the end of this chapter. Scripture Union have produced *Which Way,* which portrays the choice facing a young man who is seeking his future field of service.

Other programmes are available that give background information on the political, social and religious situations in many troubled areas of the world such as countries of South and Central America, Eastern Europe and Russia.

## Women's meetings

Ladies, have *you* ever thought of using videos in your meetings? Can you visualise a gathering in your home where the guest speaker is

Joni Eareckson, Corrie Ten Boom, Evelyn Christenson or Elisabeth Elliot? Programmes by all these women are available right now. In fact, Joni and Corrie are probably two of the most widely viewed 'video stars' by both men and women, believers and non-believers. Their powerful personal testimonies of triumph over adversities have been used time and again to bring blessing and inspiration.

Whether your women's meeting is in a home or a church hall there are programmes of special interest that you can use. Young Wives or Mothers Union will appreciate the large number of programmes aimed specifically at the Christian mother, dealing with family problems and joys, faith and prayer. You will find them stimulating to watch as well-known Christian ladies share with you their joys and difficulties in these areas. As most programmes confine themselves to around half an hour you will have ample time to follow up with brief discussions.

A four-part series from America covers *The Christian Family*, with the subtitle *A Woman's Perspective*. In the first half-hour programme Evelyn Christenson discusses 'Prayer in the Home'; in the second Elisabeth Elliot talks on 'Man's and Woman's Roles'; in the third Ann Kiemel tells how 'The Family Makes the Person'. The fourth tape is an hour long and features Dale Evans who considers 'Woman's Place in God's Plan' and shares her personal experience of losing three children in 'Overcoming Grief'.

If Mary Whitehouse is more your mark she has been recorded, at Celebration '84, talking about the work of the National Viewers and Listeners Association, and how the Lord has been leading her.

## Other home-based meetings

Using your home as a base for outreach, teaching, bible study, etc., is both rewarding and challenging. You will find yourself exposed to all sorts of people, some of whom will not share your point of view. Don't approach it negatively, but use each situation to see what God is trying to teach you (it may be that you will need to get into the word more often yourself!). Prepare and plan each activity well and ask the Holy Spirit to be a real presence at each gathering, so that folk will *know* that they have not just come to a friend's home, but they came to meet Jesus.

Some of the videos mentioned earlier are intended for viewing by a committed group. Others are suitable for a variety of mixed audiences. An excellent series for young and old Christians, produced by Lella and distributed by Trinity Video, is *Jesus Then and Now*. This twelve part series (on six cassettes) is truly a video magazine with its variety of techniques. It is hosted by David Watson and Tina Heath and uses drama, dance, cartoon, interview, documentary film and investigative reporting to transmit its message. The themes of the individual programmes are relevant contemporary issues and problems, and a variety of prominent Christian leaders from around the world appear. There is an accompanying study book.

This series was made some time before the death of David Watson. His life and attitudes are celebrated in a two-part video. *The Best is Yet to Be* is compiled from the series he made for TVS in Company. On the same tape is appended a *Tribute to David Watson* by those who knew him. This video is a good discussion starter as well as a memorial to a much-loved man.

Another useful programme for an audience that contains some who are not yet believers is *Ordinary People*. This was produced by the Christian Television Association for use in Mission England outreach and shows three sets of people and how encounter with Christ changed their lives.

An early film from the Church of Scotland dealt specifically with housegroups and their function. It is called *People Need People* and would be useful viewing if you are considering forming your own housegroup.

Although not strictly a housegroup, children in the home need looking after as well. We have two (more than once I have thought there were more!) and their needs are as real as any adults. Have you thought of using the video recorder to minister to the children? I am thinking here not of its formal use in a structured situation (that is covered thoroughly in another chapter) but of its informal application in a family setting.

In all the country's state secondary schools teachers have access to a video recorder. For many years teaching in the form of broadcast television programmes for schools has been a regular part of the curriculum. Children respond to a tv programme even better than adults. It has been a part of their lives since they were born. Also (and this places a heavier responsibility on the shoulders of producers) they are often sterner critics than adults of shoddy production standards. Think creatively about how you could use the box to share the gospel with your children.

So there you have it – an opportunity to either grasp or let slip. Making use of the video recorder in the home is, or can be, as natural as inviting a friend round for a chat and introducing them to another friend who has something important to tell them.

## Videos for teaching or bible study

| | | *duration* | *distributor* |
|---|---|---|---|
| Genesis | five tapes, each | 1:30 | Bagster |
| The Tabernacle (Billy Strachen) | | | Videout |
| The Tabernacle (Christopher Smith) | | 0:30 | ICI |
| A Mind Behind it All | three tapes, each | 0:30 | Bagster |
| The Measure of . . . | five tapes, each | 0:30 | Bagster |
| Literature Evangelism | | 0:50 | OM |
| Dynamic Christian Living | eight tapes, each | 1:00 | CWR |

## Missionary videos

| | | |
|---|---|---|
| Faces of India | 1:43 | CTA |
| Cliff in Kenya/India/Water, water | 0:56 | Tear Fund |
| Who Sends Them | 0:13 | SIM |
| The Wycliffe File | 0:45 | Wycliffe |
| People, Like Us | 0:17 | SAMS |
| Which Way | 0:24 | SU |

## Videos for women's meetings

| | | | |
|---|---|---|---|
| Reflections of His Love (the story behind *Joni*) | | 0:55 | Bagster |
| Corrie (the story behind *The Hiding Place*) | | 0:59 | Bagster |
| The Christian Family | four tapes, three of | 0:30 | Trinity |
| | one of | 1:00 | Trinity |
| Mary Whitehouse | | 1:30 | ProVision |

## Videos for discussion groups

| | | | |
|---|---|---|---|
| People Need People | five parts, totalling | 1:39 | Ch Scot |
| Jesus Then and Now | six tapes, each | 1:00 | Trinity |
| The Best is Yet to Be/David Watson: a Tribute | | 1:28 | Trinity |
| Ordinary People | | 0:35 | CTA |
| Open House | eleven discussion starters | 0:58 | CPAS |

## Some distributors

| | | |
|---|---|---|
| Bagster Video | Westbrooke House<br>76 High Street<br>Alton<br>Hants GU34 1EN | 0420 89141 |
| Christian Television Association | 27 Ash Hayes Drive<br>Nailsea<br>Bristol<br>Avon BS19 2LQ | 0272 856447 |
| Church of Scotland | 121 George Street<br>Edinburgh EH2 4YN | 031-225 5722 |
| CWR | Box 11<br>Walton-on-Thames<br>Surrey | 0932 244499 |
| Falcon | CPAS Mission at Home<br>Falcon Court<br>32 Fleet Street<br>London EC44 1DB | 01-353 0751 |
| International Correspondence Institute | PO Box 47<br>Dorking<br>Surrey RH5 5YY | 0306 711899 |
| Operation Mobilisation | PO Box 17<br>Bromley<br>Kent BR1 3NR | 01-464 9817 |
| ProVision | 5 Lindsey Road<br>Uffington<br>Stamford<br>Lincs PE9 4SH | 0780 52214 |
| Scripture Union | 130 City Road<br>London EC1V 2NJ | 01-250 1966 |

| | | |
|---|---|---|
| SIM International | 84 Beulah Hill<br>Upper Norwood<br>London SE19 3EP | 01-653 3953 |
| South American Missionary Society | Allen Gardiner House<br>Pembury Road<br>Tunbridge Wells TN2 3QU | 0892 38647 |
| Tear Fund | 11 Station Road<br>Teddington<br>Middx TW11 9AA | 01-977 9144 |
| Trinity Video | Westbrooke House<br>76 High Street<br>Alton<br>Hants GU34 1EN | 0420 89141 |
| Videout | Carrubbers' Christian Centre<br>65 High Street<br>Edinburgh | 031 667 3318 |

A selection of videos and distributors is included at the end of each subsequent chapter. For a full list see *Guide to Christian Video* (second edition, Jay Books 1986).

# 3

# The Christian life

Leslie Timmins (CTVC)

An important part of the pastoral role is to prepare people for important events in their lives like the baptism of their children or marriage, to help them cope with specific problems and difficulties – bereavement or unemployment perhaps, to encourage them to grasp opportunities, fulfill their own potential and realise the creative possibilities of every situation and the resources available to help them, and to guide them in making crucial decisions about their future.

The function of counselling is being taken seriously by the church. These areas are not the exclusive province of the theologically trained priest or minister. In fact, some people find it easier to discuss their everyday problems with others of their own age and sex, those who have faced the same difficulty or the members of a 'share and care' group who they know can be trusted. Video can help in all these areas, and many programmes have been devised to deal with specific topics.

These videos have to be suitable for, and acceptable to, a diverse audience, and it is here that we come up against a particular problem. It is simply the problem of money.

The growth of 'special interest' video has been very visible and very rapid, and commercial programmes can be bought or hired cheaply. For the odd pound or two it is possible to hire feature films, and for quite reasonable amounts special interest videos can be bought and kept for your own library.

The low price is a consequence of huge turnover and the fact that many of these programmes have already covered the cost of making them through feature-film presentation, wide television exposure, and the like.

By definition, however, the kind of programme we are talking about will be for comparatively small minority markets, and if they have had a pre-sale to television, they will have been acquired by television companies at very low rates. The result is that while the production values need to be as professional and of as a high standard as any video tape available, the cost of production means that the selling price and even the hire rate is not normally competitive with the High Street dealer.

I believe all this needs to be said because those facts spring from our experience at CTVC. We have numerous video tapes for sale and hire most of which have been (or are currently being) broadcast. Yet, because of the cost of production, the retail price has necessarily to be set at levels which are special to the market as it exists.

If, as we all hope and expect, the use of video in education by church groups and leisure and specialist groups expands, then with increased turnover the retail prices can be reduced. In the meantime, perhaps it is as well to understand that market pressures apply to this field as much as to others.

### *Programme excellence*

This leads me to look at a very important aspect of video programming, and that is quality in the offering. Just because we are talking about material made for Christian purposes, we must never allow ourselves to think that second-best will do.

Church and social groups don't mind when their own people put on the annual panto or passion play and the result is amateurish – after all everyone is doing their best and they are all in it together. So allowances are made when the public-address system fails and the Principal Boy forgets her lines!

But put the same group in front of a television set and play them a tape. Some quite definite reactions are clear. People see a lot of broadcast television. Whatever their opinions about programme content, they get highly professional technical presentation. Let them switch on a badly made amateurish video tape – and they will switch off. In my view this is a correct reaction. So the first demand is for programme excellence.

### *Programme content*

For the purposes of this chapter, I would like to apply three questions to the contents of the videos, in order to define their applications.

1. *Is the production meant to be complete in itself?*

If so, it may be intended simply to offer information on social or domestic situations, or be some form of documentary showing interesting activities or people.

2. *Is the production likely to provoke discussion?*

It is then used as a 'tool' for housegroups, preferably accompanied by a list of questions for discussion and presented by a group leader who does not see the video tape as an end in itself but as stimulus for people to ask their own questions and produce their own responses.

3. *Is the production meant to stimulate people into action?*

It is then used as a brief for people to work out appropriate action for themselves or their own group.

I take these three sets of usage as representing common and wide uses of video tape and I make no apology for using CTVC productions as examples.

## The complete programme

Although, of course, this type of programme is not intended to preclude discussion, it can be complete in itself because it has an entertainment element as well as its more serious intention. In passing, I might say that the entertainment element is very important because if your audience, however small, is not gripped and even to some extent entranced, the message will not reach them half so effectively.

One example of a complete programme is *Yvonne Hudson: Sculptor*. This is a reflective and thought-provoking profile of a farmer's wife living in Chichester. Primarily a sculptor, now working mainly in ceramics, Yvonne believes that new skills and ideas should continually be developed; she took up ecclesiastical embroidery six years ago, and is now developing a form of tapestry-weaving.

Much of her work has a religious basis, and can be viewed in local churches and Chichester Cathedral, but Yvonne surprisingly describes Sunday as the 'least day'. She feels that a living faith is ". . . the way you live, the way you treat other people, the way you get on with other people . . . these are the things that contribute to a real faith".

*The Weekenders* is a description of the Methodist Association of Youth Clubs' London Weekend, which is the biggest Christian rally for young people in Europe, and has been held annually since 1945. It is the highlight of the year for young Methodists, who not only travel from all over Britain to take part in events, but also spend several months preparing for them.

This documentary is a fast-moving production that captures the mood and excitement of the weekend, and gives an idea of some of the backstage planning that makes each year seem better than the last.

The show itself is held in the Royal Albert Hall, but is rehearsed by different groups throughout the country. Even the spectacular finale with its cast of over a thousand is rehearsed in segments in village halls and car parks. After a brief run-through it's 'on with the show' and audience enthusiasm is even higher than the last night of the Proms.

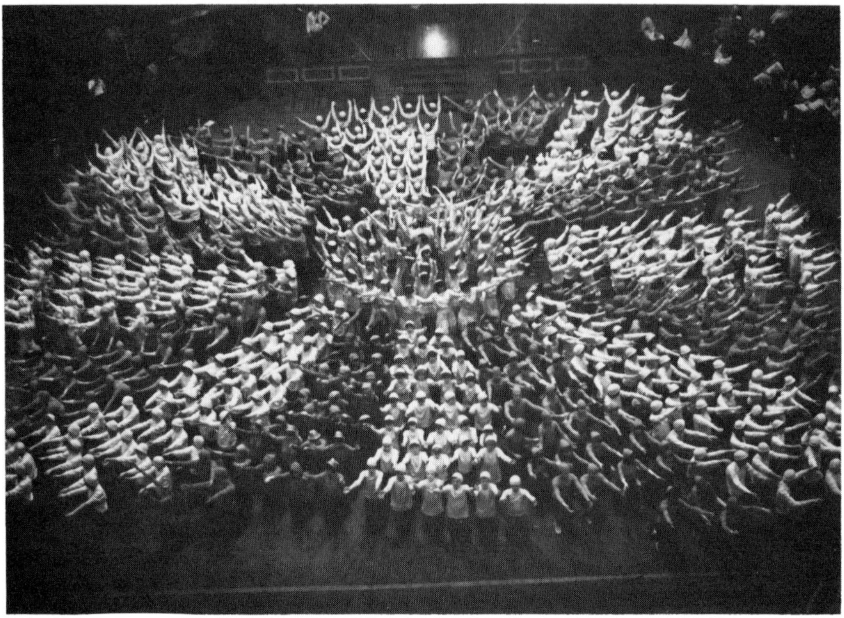

*The spectacular finale in* The Weekenders *(CTVC)*

Meanwhile at Westminster Central Hall, the weekend's main music event, Tempo, packs the 3000 seat auditorium. The music is mostly loud and the beat heavy, but the lyrics are Christian, and not one of those present remains unmoved by the excitement of the weekend.

Sunday begins with worship in the Albert Hall, and for many this is the high spot of the weekend. The service is so popular that the organisers have to run it through twice, each time with a congregation of 6000. The weekend, and the film, ends with a march from Central Hall, Westminster, to the mass rally in Trafalgar Square.

A contrast is *Marilyn Baker: Songwriter.* Marilyn is a blind music teacher who believes that God uses her musical talents to communicate his word through song. She describes herself as 'ordinary' but the programme shows that this is far from the case. She teaches the oboe, the clarinet and the piano, and the film is as much an insight into what it is like to be blind as it is to have faith.

In the film Marilyn tells how God gives her the ideas for the songs, both words and music. Her faith is direct, unfaltering and simply expressed through song, and the message is of love and hope.

I have described these three programmes at some length because I believe they provide an excellent example of how to use video in a housegroup as a means of showing how the Christian life can be lived and how the Christian faith sustains people.

This could be the kind of group where the deliberate policy is to invite people on the fringe of church groups to an entertaining and interesting evening with a chance to go a little deeper as the talk develops afterwards.

I regard this use of video as being 'pre-evangelistic'. Programmes like those described are unlikely to convert a viewer by themselves, but they can be used effectively to give a leader or counsellor an excellent opportunity to lead on from exciting or moving or reflective Christian experience to the needs of those willing to go a step further. They also give a useful insight into the nature of a real Christian life that can benefit nominal church attenders.

The value of such documentary programmes cannot be over-stressed, because there is undoubtedly a large number of people who are enquiring, asking questions, wondering about the quality of life today, and the meaning of their own lives. They do not seek answers or responses in the conventional framework of worship but may well be drawn to the faith by such informed methods.

## The discussion programme

This type of programme will be likely to be intended for a rather more committed group who may want to examine crucial questions at length. An example is *Happy Birthday Comrade Martin.*

Olympic gold medals and the Berlin Wall is the popular image of East Germany today. But 500 years after Martin Luther was born, in what is now a Marxist state, the protestant church has not only survived, it is allowed to play a key role in the life of this communist society.

Paul Oestreicher returned to the land where he was born to report on how the protestant church that Luther founded is coping with the realities of a communist state. It can be seen that although the state imposes clear limits on what the church is allowed to do, both sides tend to avoid open conflict. An uneasy detente has been arrived at as both sides have recognised their common enemy – the growing apathy towards both religion and Marxism in East Germany's consumer society, influenced as it is by its blood brothers of Western Germany.

Oestreicher, International Secretary for the British Council of Churches obtained exclusive interviews with Klaus Gysi, Secretary of State for Church Affairs, and Bishop Johannes Hempel, Primate of East Germany. He also looks at life styles in a country village and a

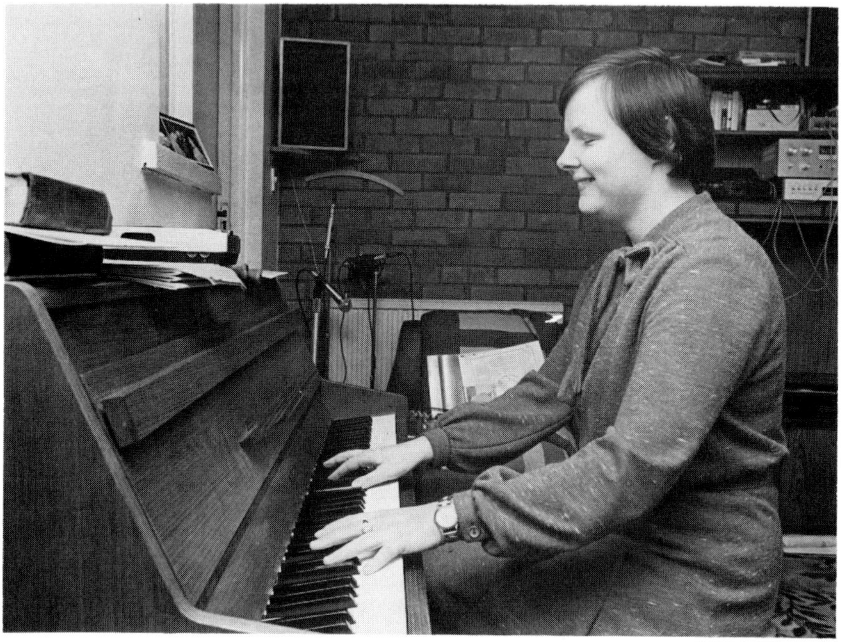

*Marilyn Baker is a blind music teacher who tells how God gives her the ideas for her songs, both words and music, in* Marilyn Baker: Songwriter *(CTVC)*

new town, and visits a home for the mentally handicapped run by the church in East Berlin and a church hospital in Dresden which was rebuilt with the aid of British students from Coventry after most of the original had been destroyed by Anglo-American bombing.

*Mission Then and Now* is a four-part bible study in the Acts of the Apostles introducing men and women who followed Jesus and found that the resurrection transformed their lives. It indicates how the church can rediscover this vitality. The studies are by the Rev Dr Donald English, a distinguished preacher and evangelist.

It is in sharp contrast to *Happy Birthday Comrade Martin.* Paul Oestreicher raises important social, political, and international questions. This type of programme should be a way of widening the vision and thought of people who can often be too concerned with the 'parish pump' and it certainly will provoke lively discussion.

*Mission Then and Now*, on the other hand, contains traditional bible studies by a distinguished scholar and preacher. Other bible study programmes have been mentioned and listed in the previous chapter. At a time when the theological debate is so public and sharp it seems very important to provide straightforward teaching which is still susceptible to discussion so that the essential 'food for thought' is provided.

This is a more pastoral use of video in the teaching and educational fields.

## The action programme

A third use of video may be the programme which leads to definite action by the people who see it. A good example is *Nuts and Bolts of CARE*.

In communities both large and small, all over the country, volunteers are getting together in organised groups to help those who are unable to help themselves. This half-hour documentary shows how just such a scheme was created and now operates successfully in Guildford, Surrey. It is a step-by-step guide covering everything from the initial setting up and organisation to the publicity and the type of service offered.

CARE schemes provide 'alternative' care for those people who cannot be dealt with by the overworked social services or their local churches and whose families or neighbours are not able or willing to give up their time. Volunteers who work in CARE schemes offer only the kind of help they would like to give, and are called upon as often or as little as they would like. Those with a lot of spare time may have one or more calls a week, whereas those with less time may only be called once every six months.

Some people volunteer their cars for visits to the shops, transport to hospital, or just for 'drives' in the countryside; others may visit the sick or lonely, relieve relatives of watching over the elderly or handicapped or look after small children whilst their mums have a morning off; some volunteer their gardening skills, hang wallpaper and paint, or simply moving heavy furniture for those who cannot manage it themselves.

This programme is chosen because to my certain knowledge a large number of CARE schemes have been set up either in direct response to, or through using as a 'starter tool', the programme or its predecessor, *Time to Care.*

The programme comes backed up by literature to help groups to start their own scheme, and the results speak for themselves.

In the same way *Retirement Report* has been of use in helping people and retirement training groups to face the problems of retirement.

This kind of practical outcome of a piece of video group work is intensely satisfying and can often produce concrete and visible results in concerned social action.

A special kind of action programme is that which is intended to advise and guide people who have already been considering some sort of action, or are having to face some inevitable experience. Such programmes are more limited, but warrant a heading of their own:

## Special programmes

Under this heading we can consider specialist video productions about instruction on such subjects as marriage, baptism and the sacrament.

A difficulty which, frankly, I have not been able to overcome in these very specific areas is that of the wide divergence of practice.

None of these subjects can be dealt with ecumenically because different denominational practices and beliefs vary so widely. But even within the same denomination there is such a widely varied number of attitudes that the task becomes complex.

For example, it would seem on the face of it to be a simple matter to produce a video programme to support a priest or minister in preparation for marriage. Simple, until we take a look at attitudes about re-marriage of divorced people, the questions which are raised about who should be able to marry in church, the differing views held by those who are asked to officiate.

The same enormous differences emerge when we look at the possibility of making an instructional programme about the Holy Communion. Is that what we should call it? Or should we speak of the Eucharist or The Lord's Supper or The Mass?

I need say little more to illustrate the problem. It is possibly here that the do-it-yourself home video enthusiasts might well come into their own. Certainly the potential viewer will be concerned to ensure that a programme accords with his own attitudes before hiring or buying it (although we might all benefit from occasionally studying the practices of other Christian groups).

One attempt we have made at CTVC to touch upon the questions of marriage (and divorce) is *not* an instructional programme, but rather investigates the emotional price paid by people involved in divorce. *Till Divorce Do Us Part?* looks at seven people. In every case the realities of divorce came as a total and often daunting surprise. No one had any idea of what to expect. They were as ill-prepared for the experience of divorce as they were for marriage in the first place. It can take years to recover.

But eventually most people do recover and what happens can often be the biggest surprise of all. "I've got faith in myself now", says Moyra who is bringing up two children on her own, "and in my capabilities. I've even amazed myself at some of the things I've done . . . I'm a much happier person, much more relaxed".

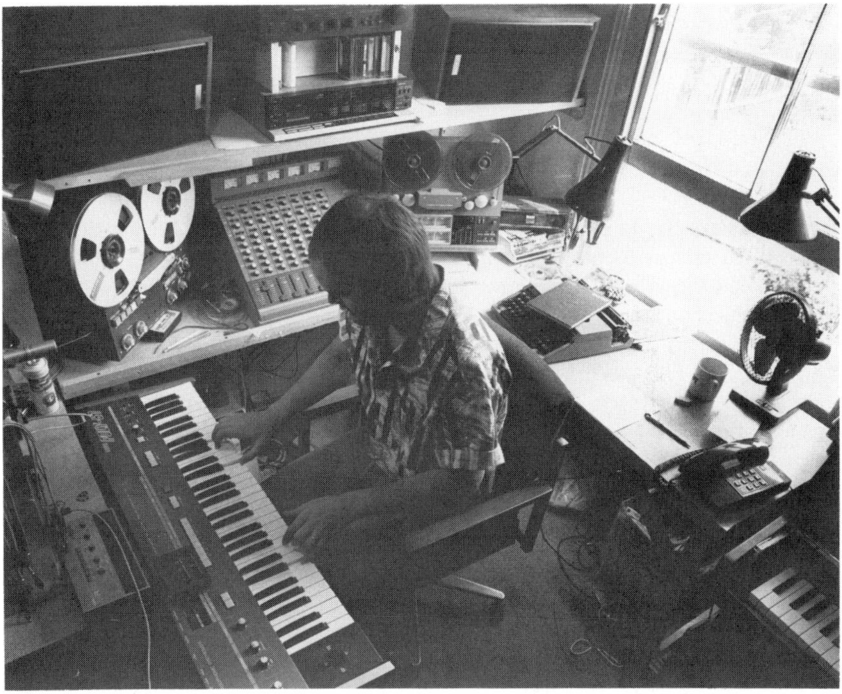

*Paul Field, with 15 years in the contemporary music business and six albums behind him, is one of the most successful writers/performers in today's Christian popular music scene. The documentary video takes its title from one of his compositions* Rescued for a Reason *(CTVC)*

This programme has much to say to anyone contemplating divorce, to anyone recovering from it, and to anyone with a divorced or separated person in the community. The irony is, it has even more to say to anyone contemplating marriage.

Similar work is available on the important subject of bereavement. It seems clear to me still that there is a kind of pornography of death. People are not encouraged or helped to grieve properly. The 'sanitisation' of death with the modern methods of 'protecting' relatives from being present at the moment of death and keeping the whole thing somewhat at arm's length is still very much a problem and we have tried to face that in our productions.

Under the heading of 'special programmes' might come something like *Responding With Joy*, a video made by CTVC to show what effective stewardship can be, and what effect it can have on mission.

When all is said about video use of any kind, however, it can never be a substitute for good leadership. It can be a first-class aid. Well presented, attractively offered, it is a new and effective way of opening doors for people who may never seriously consider a more formal contract with the church, and it can reinforce the faith and effort of those already committed.

## The authority of video

I would like to express a last thought on the nature and content of Christian communication and what we call the *mass media.*

It takes the form of a question. Is it true that just because a very large number of people hear a message simultaneously, this makes the statement more true than it otherwise may be?

"Millions of people are listening to what I am saying at this moment", declares the 'electronic preacher' dramatically. Is the implication that therefore he carries more authority and therefore more truth than others – that in some way he is, indeed, more effective than others?

There is, I believe, a danger that such an assumption is too easily made.

Of course, home video does not, and probably will not, command the audiences imagined in millions that can be reached by broadcast television. However, a well-made and effective video tape will carry the same message to a listening audience which largely outnumbers the audience receiving that message from that preacher or teacher in his own pulpit or classroom.

Twin dangers present themselves. The first is to fall into the trap of believing that because there is large, almost simultaneous, audience the message itself becomes more important, more powerful, even, in some mysterious way more 'true' – whatever that means.

The fact is that the message only has the importance, power or truth that it carries within itself. The communicator's self-criticism, his constant need to examine his own integrity and his willingness to test the authority of his message against the teaching of the Bible and the received wisdom of the saints, is therefore imperative.

The glib, well-tested tricks of oratory, the oft used, well turned phrase which has become a cliché, has no place here. Newly minted and fresh language and image may be more necessary, and searching thought must support this kind of communication.

The second danger is to look towards mass communication as a replacement for the essential 'one-to-one' dialogue which is at the heart of communicating the Christian faith.

So, at the risk of overdoing the repetition, let us re-emphasise that the video tape should be regarded as a tool. It can never take the place of shared personal experience and fellowship. It can be an introduction, an attention-taker, a form of instruction and education.

But in the last analysis the ultimate mode of Christian communication will be personal and corporate.

## Complete programmes

| | *duration* | *distributor* |
|---|---|---|
| Yvonne Hudson: Sculptor | 0:29 | CTVC |
| The Weekenders | 0:28 | CTVC |
| Marilyn Baker: Songwriter | 0:28 | CTVC |
| Rescued for a Reason | 0:28 | CTVC |
| The Invisible Men on the Jesus Beat | 0:26 | CTVC |
| Janet and Mysia | 0:28 | CTVC |
| Sylvia Read and William Fry | 0:29 | CTVC |
| Sing to the Lord | 0:26 | CTVC |
| Caribbean Partnership | 0:25 | Ch Scot |
| Christmas in Shanghai | 0:30 | CFWM |
| Jesus and the Fishermen | 0:27 | Sussex |
| The Pope in the Park | 0:54 | Masterclass |
| Christians in Sport | 0:30 | Christians in Sport |

## Discussion programmes

| | | |
|---|---|---|
| Happy Birthday Comrade Martin | 0:52 | CTVC |
| A Man Called Luther | 0:40 | Bagster |
| Mission Then and Now | 1:20 | CTVC |
| Mission Impossible | 0:47 | Meth H M/CTVC |
| The Silent Witness | 0:55 | Screenpro |

## Action programmes

| | | |
|---|---|---|
| Nuts and Bolts of CARE | 0:24 | CTVC |
| Retirement Report | 0:29 | CTVC |
| Root Groups | 0:14 | USPG |

## Special programmes

| | | |
|---|---|---|
| Till Divorce Do Us Part? | 0:29 | CTVC |
| Responding with Joy | 0:27 | CTVC |
| The Italian Connection | 0:27 | CTVC |
| Back to Square One | 0:29 | CTVC |
| Fixed for the Future | 0:28 | CTVC |
| The Life that's Left | 0:25 | CTVC |
| Stairway – a Parents' Guide to Baptism | 0:15 | Falcon |
| Choices | 1:02 | Church Army |

## Some distributors

| | | |
|---|---|---|
| Bagster Video | Westbrooke House<br>76 High Street<br>Alton<br>Hants GU34 1EN | 0420 89141 |
| Christians in Sport | PO Box 93<br>Oxford OX1 1QX | 0865 240407 |
| Church Army | Audio-Visual Resources Dept<br>Independents Road<br>Blackheath<br>London SE3 9LG | 01-318 1226 |
| Church of Scotland | 121 George Street<br>Edinburgh EH2 4YN | 031-225 5722 |
| Conference for World Mission | 2 Eaton Gate<br>London SW1 9BL | 01-730 9611 |
| CTVC | Beeson's Yard<br>Bury Lane<br>Rickmansworth<br>Herts WD3 1DS | 0923 777933 |
| Falcon | Church Pastoral Aid Society<br>Mission at Home<br>Falcon Court<br>32 Fleet Street<br>London EC4Y 1DB | 01-353 0751 |
| Masterclass | Holiday Brothers<br>172 Finney Lane<br>Heald Green<br>Cheshire SK8 3PU | 061-437 0538 |
| Methodist Church Home Mission | 1 Central Buildings<br>Westminster<br>London SW1H 9NU | 01-222 4776 |

| | | |
|---|---|---|
| Screenpro Films | 5 Meard Street<br>London W1V 3HQ | 01-439 7691 |
| Sussex Video | Townsend<br>Poulshot<br>Devizes<br>Wilts SN10 1SD | 038 082 337 |
| The United Society for the Propagation of the Gospel | 15 Tufton Street<br>London SW1P 3QQ | 01-222 4222 |

# 4

# Training for service

David G Hamilton (Church of Scotland Education)

One of the least exploited applications of video so far has been in its use as an aid in training or educating Christians to be better qualified in whatever role they are called upon to play, be it in the community as a lay visitor or playgroup leader, in the church as a Sunday school teacher or housegroup leader, or in a variety of jobs in the foreign mission field.

## Television as an educational resource

From its inception television has been recognised as having an enormous potential as an educational resource. This potential is still far from being realised but there are many encouraging signs of the coming 'new day'. The remarkable progress of developed educational services, such as American Public Television schools programmes or the British Open University, indicates something of the truly impressive possibilities of television as an educational medium.

Educators within the television companies have been quick to seize upon the best of modern educational theory and practice and probably, wittingly or otherwise, have done much to win widespread credence for such approaches. For example, the language 'lesson' as presented on schools television is typically a half hour 'spent' in a foreign town, listening to people speak, while we watch them doing things and meeting each other, in a total social context. It is as though we are there. We are invited into the experience. Through caption (the electronic equivalant of 'flashcard' display), through action replay, through carefully written scripts which are often humorous, the familiar methods of experiential learning, reflection, extension, revision and evaluation are given a dimension not otherwise attainable.

In terms of educational technology the great advance is in the realm of 'education by seeing'. The visual demonstration, whether to do with pruning roses or making a giant cardboard Goliath for the Sunday school class, represents an enormous step forward in mass communication. Much of this 'learning by seeing' is of the essence of television programmes generally.

It is useful here to recognise a distinction between programmes that are *educative* inasmuch as they inform and stimulate further reflection and interest and those programmes which are by specific design *educational*, developed to fulfil clearly conceived aims and objects in the context of a formal learning programme. David Attenborough's *Life on Earth* was educative, but then so also was the popular *Upstairs, Downstairs*, with its contrast in social conditions and portrayal of social change in early twentieth century Britain. A clear example of an educational programme is the schools broadcast *Maths Counts* (BBC2), but so also is the peak evening viewing programme on the history of Scotland, *Scotland's Story* (C4).

BBC Scotland has for some years screened *The Beechgrove Garden*, a programme full of instructions, suggestions, useful tips and information, presented by two gardening experts with a delightful and pawky sense of humour. Some might argue as to whether it is truly an educational programme. It is directive in its educational intent yet good entertainment. Like such classics as *Sesame Street*, it serves to remind us that learning can be fun – and this is a dimension in which television excels. The John Cleese management training videos have shown just how effectively entertainment/humour and learning can be combined.

## Television/video in church education

The use of video as an educational/training medium in the church is still in its infancy. Few congregations have yet absorbed television/video into their battery of resource aids as an accepted element in their regular programmes and ministry. Most congregations who use video do so on an occasional, almost tentative, basis. Many congregations either fail to see the need for this medium or are convinced that they can get by without it perfectly well.

Brian Brown's important study of the attitudes of church people to television in general and video in particular, although based upon the restricted survey of ministerial and lay leadership within the Methodist Church, most probably reflects the situation across the denominational spectrum. This research suggests that congregational leaders are largely unaware of the ways in which their ministries might be advanced through this medium. Some regard the necessary capital investment as being beyond their financial resources, others regard it as simply not being a priority, and just as many appear to be apathetic to the possibilities of video. In addition there are some who, with some justification, await the availability of improved equipment such as good quality large screens suited for use in large halls and of sufficient video programmes of quality to justify the capital expense. It is interesting to note that only 13 per cent of the church members surveyed either owned or rented a video recorder compared with the national penetration figure then of 26 per cent. In a word, Christian leaders are less persuaded of the worth of video than is society as a whole.

This raises serious questions about communicating the advantages of television as a resource and the growing importance television will play in education in the years ahead. It also raises questions about the training strategies and techniques used by churches generally and perhaps even exposes fundamental weaknesses such as an unreadiness to undergo systematic training and preparation. In Scotland, after decades of frustrated efforts to encourage and promote lay training, particularly among elders, we are now witnessing some very heartening developments with a marked increase in response to training courses, and *significantly*, these courses are now resourced by a series of video programmes. The same also is true of Sunday school teaching, and again of stewardship visitation. Nevertheless, those lay people responding to training opportunities remain a small if vitally important minority. The Church not only has great need of enthusiastic and committed Christian activists but must expect those who volunteer for particular avenues of service to undergo preparation and training for their tasks. So fundamental is this that we take a moment to reflect on its significance.

## Training for service

The recovery of the doctrine of the priesthood of all believers in our time has given fresh and welcome impetus to our ideas on Christian stewardship. There is in the modern Church a new readiness to develop our God-given talents in the service of the Lord. We have a new awareness that in being called to serve we are also charged with the responsibility of doing what we do for God to the best of our ability. At the close of Sunday worship I frequently discharge the congregation with the words:

"Your service here is ended: your service in the world continues. Go therefore in the strength of the Lord . . ."

Behind those words is a whole theology of Christian service in which our service of worship and the service rendered in our daily living is part of the one offering we make to the glory of God. In this sense the ideas of worship and service are interchangeable. Nothing but the best is good enough as we praise God with our lives.

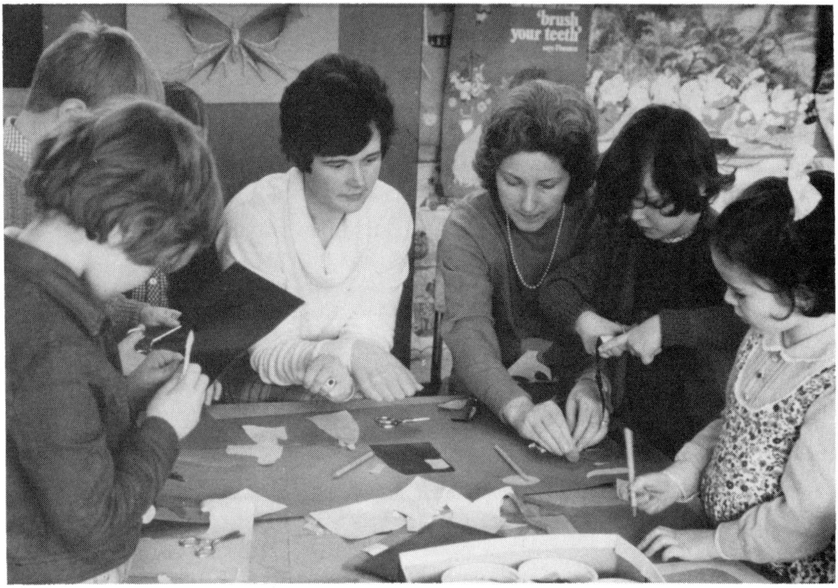


*Sunday school teachers with an activity group in* Help Me to Teach! *(Church of Scotland)*

Another concept central to the idea of the priesthood of all believers is that of particular gifts. This recognises that in Christian service we are called upon not only to serve Christ corporately as his body but also to express our individuality through our particular talents, personalities, aptitudes and interests. We find our respective vocations in doing what we do best to the glory of God.

Taken together these basic principles of Christian service encourage us not only to use our God-given talents but to develop them, extend them and strive for excellence in the exercise of them. In plain words, Christians should be good at what they do. They should be efficient and effective. They have the highest possible motivation for *developing* and *protecting* their skills.

So it is also with those engaged in teaching and training and those whose task it is to plan the training opportunities and develop the necessary resources. The educators, more than any others, must be open to new methods, better ways, developments in educational technology. This is our vocation – to help Christians be the Church, to fulfil the vision of the kingdom.

## The task of the Christian educator

The Christian educator, every bit as much as any other teacher, needs to be very clear as to whom he is teaching, why he is teaching them, what he is teaching and how he will go about the task. Not only does he require to know his subject, he must know his people and be able to encourage their learning in a programme that is stimulating and which achieves prescribed targets. It follows from this that in the whole field of Christian education resources the educator is *himself* the principal resource. Even in the new age of information technology, and perhaps particularly in such an age, the most important factor is the person who controls the learning session. The ability of the leader, or teacher, or trainer to use resource materials in a way that is appropriate to his aims is the hallmark of the effective educator. This is true in the use of published syllabuses, scripture portions, films, overhead projectors, flipcharts and pictures – just as it is of television and video.

This is a point of immense importance. While it is already clear that video programmes constitute an invaluable resource, they remain simply that – a resource, nothing less and nothing more. The leader or tutor has a responsibility for using video as a means of fulfilling his aims and objectives. The mechanics of operating the video recorder are quickly mastered. The purposes for which video will be used in education require rather more careful reflection. The unfortunate fact is that video can be used in ways that are frankly counter productive.

I have seen groups goggle at my Sunday school teacher training video, *Help Me to Teach!*, right the way through all six units of the 65 minute tape without a break, as though it were a John Wayne western, and surface at the end, groggy and saturated with ideas, complaining that "it was all a bit much"! In such cases the teacher has abdicated. He has handed over his entire function to the box. He has rendered himself redundant.

I have also seen gadget-conscious leaders fiddle and fidget with the video/receiver to an infuriating extent, adjusting volume, brightness, contrast in turn; punching the *pause* button in order to expound at length on a trivial detail alluded to in the programme; pushing the *fast forward* button to hurry on to 'the next important bit' – and all the while frustrating the very process of communication he is there to foster. In both instances the misuse of video impeded learning.

Clearly it is important that, as with all resources, the teacher should be master of the video programme and not a slave to it. He should be prepared to select from the mass of material available and even from within programmes. For example, I have found it more beneficial to use brief clips from our own Church of Scotland programmes such as *Meet Cliff Richard*, focusing on one or two simple points pertinent to the theme being developed rather than present the entire half-hour

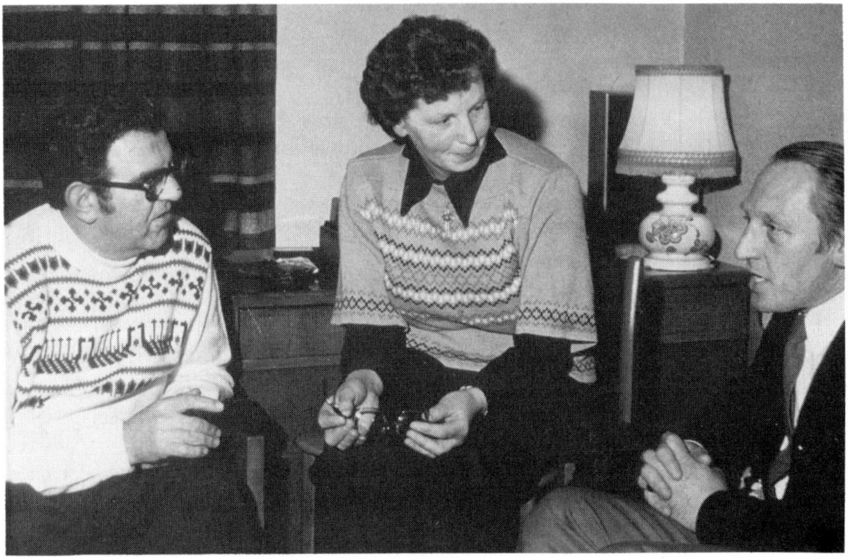

*A housegroup discussison is portrayed in* Fighting Talk *(Church of Scotland)*

free ranging discussion and seek thereafter to pick up a number of points as recalled by the group. The latter approach I find to be too rich in content, too diverse and free ranging in themes, almost to the point of being indigestible. However, to offer input on what Cliff has to say about rock gospel music as one brief contribution to a presentation on worship can be of considerable value.

This approach requires of the leader a readiness to look upon many of the more general video programmes as a kind of data bank and to be prepared to sift and search for the apposite clip. Similarly, it is possible to delve into other programmes on a more specific theme with a view to presenting not the entire video but a representative point. Ian Gall's award-winning video, *The Very Gate of Heaven* is a most useful study of the Church in the Caribbean. It also offers specific material on problems of coping with poverty in developing countries.

In *Meet Dr. Allan Boesak,* the complex problems of apartheid in South Africa today are explored in an important dialogue. Even so, I have taken a clip from that video in which Allan Boesak makes a magnificent and compelling witness to the power of Christ to triumph over all adversity and used it in a quite different and more general setting as one of the most inspirational affirmations of the gospel I have experienced on the small screen. These are examples of how the educator should seek always to design his presentation with care, using only what he needs, in a way and at a point that best suits his purpose. The possibilities for this approach are limitless.

## Video training programmes

Whereas much educational video promotes reflection and discussion of an open-ended nature I have found this medium of greatest value in the training of church leadership. In the Scottish church scene video is presently being used for in-service training of ministers, the preparation of lay visitors, particularly in relation to stewardship visitation, the ongoing training of elders, the systematic training of Sunday school teachers and the training of laity in the management of small groups. Many of the programmes are of a highly specialised nature. For example, *All God's People* is a 30 minute programme on the Christian nurture of mentally handicapped children and adults. It is used with supporting literature for in-service courses for clergy engaged in chaplaincies to schools and units for the mentally handicapped. It will, however, have a second and equally important application in the raising of awareness within congregations of the possibilities open to church people for support of the mentally handicapped and their families. Another pack of literature is designed for use with church study groups.

Since video production is costly in terms of time, manpower and equipment, we endeavour to target several related areas of application for programmes of a highly specialised nature. A recently completed video, *Choosing The Way,* was conceived for use with young people in their mid-teens with a view to exploring with them the questions raised by teenagers concerning the Church. We had in mind the use of the four-part video as a resource for a five-day residential conference. We also designed the programme as a supplementary resource for confirmation classes with this age group. Finally, we saw the possibilities for using this teens-orientated programme in youth leadership programmes. These user groups were kept in mind during production and I believe the video has successfully met the criteria set by us. What is interesting is how the video is being used for associated purposes such as we had never contemplated.

Training is essentially the encouragement of the development of insights, attitudes, aptitudes and skills to the point where they might be efficiently brought to bear on the pursuit and accomplishment of specific tasks. Training individuals for service incorporates both the development of a basic understanding of the task to be undertaken and the practice of techniques associated with the achievement of the task. Both aspects may be admirably promoted using video resources. For example, in the training of Sunday school teachers, the video *Help Me to Teach!* is arranged in six units, the first of these being concerned with the question of children and their relationship to the Church. Only within this fundamental perspective is it legitimate to ask the title question for the unit, 'What's the Use of Sunday School?'.

The unit depicts churches, children in church, Sunday schools in operation, teachers obviously coping well in real life situations (rather than studio sets) giving the conviction that such a well organised possibility is achievable. The initial credibility barrier is broken down by the visual demonstration. Thereafter, in succeeding units it is possible to offer a stream of advice and useful hints along the lines of pupil control, class organisation, preparation and presentation of the lesson, the use of craft work, drama and music, all the while interspersing basic theory with ample demonstration.

In all that has been said so far about the use of video in education there has been the clear affirmation that video is one useful resource among others. It is unlikely that television/video on its own would be used as an effective means of education. There are obvious problems, such as the technical difficulties of clarification and questioning. The days of truly interactive television are some way off yet. Until then there is need for someone on the spot or available by regular arrangement who can clarify or reinforce issues, deal with doubts, misunderstandings, critiques and so on. Left to itself the television could be regarded as the ultimate in authoritarian presentation!

There is also the continuing difficulty of making the student's work active. The student needs to work through exercises or practise the skills described, demonstrated and advocated. He has to become proficient in himself. The whole process of application/assimilation of what is to be learned is obviously best achieved under the supervision of an enabling tutor. It is significant that these difficulties have been best met when television has not attempted to exhaust the teaching process, but has been offered as an aid, *in planned relation* to the work of leaders and tutors, or to other prescribed resource materials (as in the printed units of the Open University).

This interrelationship and interdependence of resource person, television/video input and the printed word is of major importance.

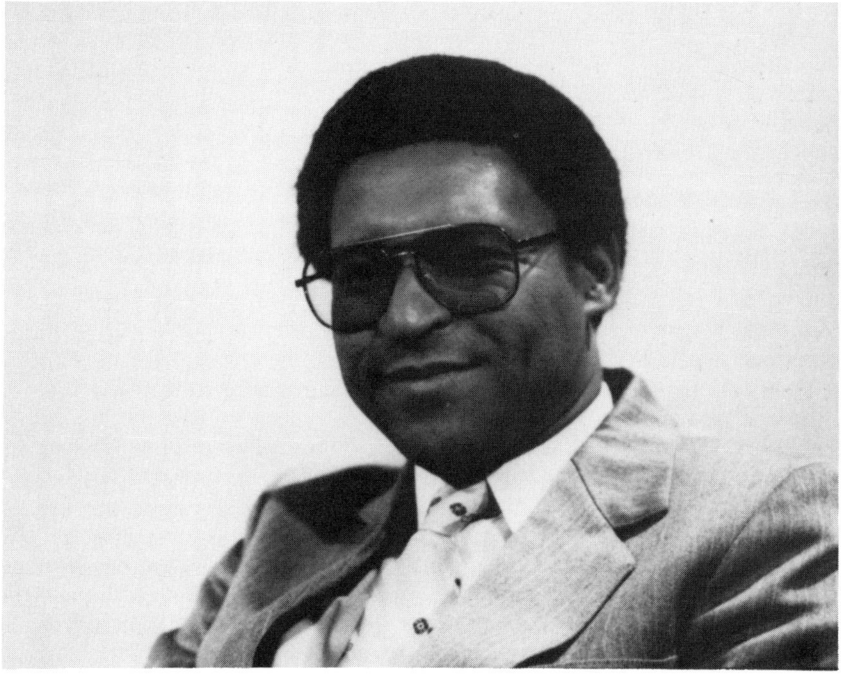

*Dr Allan Boesak, pastor, preacher, theologian and international speaker talks about his struggle for justice and human dignity for his people in* Meet . . . Dr Allan Boesak *(Church of Scotland)*

In the earliest days of educational television there were some in the publishing industry and bookselling trade who feared for the future of the printed word. The fact is that television generates books! This is true whether of the soap opera, the naturalist documentary or the sports programme. It is most certainly true of education where the printed word is very much more than a 'fixed record' of a programme enjoyed or which stimulated interest but is an inbuilt part of the total learning experience.

In the early days of Church of Scotland Video, programmes were devised with the intention that they be self-sufficient. When the Church's educators began to commission programmes this remained the basis of our thinking. I believe we have moved slowly but surely toward the concept of total package. The *Jesus Then and Now* series promoted jointly by Trinity Video and Lion Publishing did much to develop this partnership in Christian video. Whereas with *Help Me to Teach!* I adopted an open-ended inductive approach to the possibility of a multiple resource approach to a particular training venture, I would not hesitate now to plan an integrated curriculum from the outset and to conceive of video within the framework of 'resource battery'. For an effective learning programme, there has to be contextualisation, revision, reflection and follow-up through extension. Leaders' notes are to my mind an essential element in all this, and if possible, there should be student packs with fact sheets, assignments and further resource references.

## Strategies in using video as a training resource

Earlier, I warned of the extremes both of using too much video content in any one lesson and of interfering too much with the flow of the video presentation. Let us now consider some rudimentary steps toward getting the best out of video.

1. *Preview your material.* It seems preposterously obvious to say so, yet experience shows that not infrequently the presenter does not see the programme he has chosen before the actual public screening. Often he will select the programme from the catalogue with only the 'blurb' to guide him. In introducing the piece he will distance himself from the programme with some such preface as "I haven't seen this tape myself, but it is recommended", thereby leaving the way free to join in the criticism of the programme if it doesn't measure up to expectations. Let us be clear that as educators/presenters it is our job to know what's in the programme and to show only what is pertinent, helpful and pursuant of our aims. Take every opportunity to build up a working knowledge of the contents of as many video programmes as possible. A few videos will be purpose-made as specialist training resources. Many more will be of a general nature yet yield clips or extracts which are precisely what the trainer is looking for. Build up your own private catalogue by recording on index cards the salient features of tapes previewed by you. You may find it more useful to run a stop-watch on the tape rather than depend on the recorder counter for locating themes or extracts for possible future use.

2. *Select only what you need.* This is where ability and experience as an educator/leader is so invaluable. Be clear about your aims and objectives and ask whether a particular piece does actually foster the advancement of these. You cannot make clear to others what is not clear to yourself! Hold to your essential purposes. The video resource which supports, demonstrates or clarifies the point you seek to make can be enormously effective. However, do not use video for its own sake or just because it is available! If you cannot find an appropriate clip then it would be better to use another resource altogether. Having said this, there is already an extensive range of material,

religious and secular, specialised and general, readily available. Half the battle is knowing of its existence, and this can be largely overcome through obtaining catalogues, taking regular magazines, such as AVA, Christian Update or Lookout, and browsing/viewing wherever possible. The development of Christian resource centres across the nation has opened up important new opportunities in this respect, although in many cases, resource centres, bookshops and video libraries could still offer much more in the way of preview facilities. It is in the very nature of education that the educator will require to see in advance what is in a programme before incorporating it into his presentation.

3. *Set up your excerpt(s) in advance.* As with film or slide projection and sound cassettes, fiddling around with the hardware while your audience waits patiently (?), really isn't good enough. Wherever possible, have your clip pre-set, ready to run. Video recorders are so simple to operate that this is no great chore. The important thing is that you should avoid interruption of the learning session and preserve continuity. You should lead into your video presentation simply and naturally, and just as clearly round-off the extract at the conclusion of the screening.

4. *Keep your extracts reasonably brief.* In a training session, you are demanding of your audience a high degree of attention and concentration. The ability of most of us to sustain such concentration is limited. This is not a problem where you are 'splicing' into your programme video clips each of two or three minutes duration. It is, however, a real problem with longer training programmes. In my own productions I very deliberately organise the material into units or modules of between eight and 15 minutes each, partly to delineate specific areas for study and discussion, but also to guard against viewer fatigue. Certainly, I would advocate 15 minutes of concentrated viewing as the maximum limit. When using *Help Me to Teach!* with ministers and Sunday school teachers, I frequently use a ten minute video module in several 'bursts', interspersed with comment, feedback from the group and practical exercises such as poster painting, dramatic role play, story-telling. Indeed, used in this way that particular tape, organised in six units and with total tape duration of just over an hour could serve a training course of upwards of twelve one-hour sessions. Do remember training with video is not at all like watching the John Wayne movie!

5. Remember that with the video recorder you have the facility both to re-run and to freeze your programmes. Used to excess this becomes a distraction. However, there will be times when having screened a video tape or clip and having discussed it you may wish to view it again. For example, if having viewed a unit on how to communicate with children, the group then engages in practical

exercises, it is often most illuminating to share experiences and then to attend once more to the principles outlined in the video. The *pause* button on the recorder will hold a frame stationary to permit brief amplified comment by the presenter. Sports commentators use this technique, for example, to examine an 'offside' decision at a soccer match. I too have used the 'freeze' frame to identify 'offences' such as the physical layout of a group where eye contact and group dynamics matter a great deal.

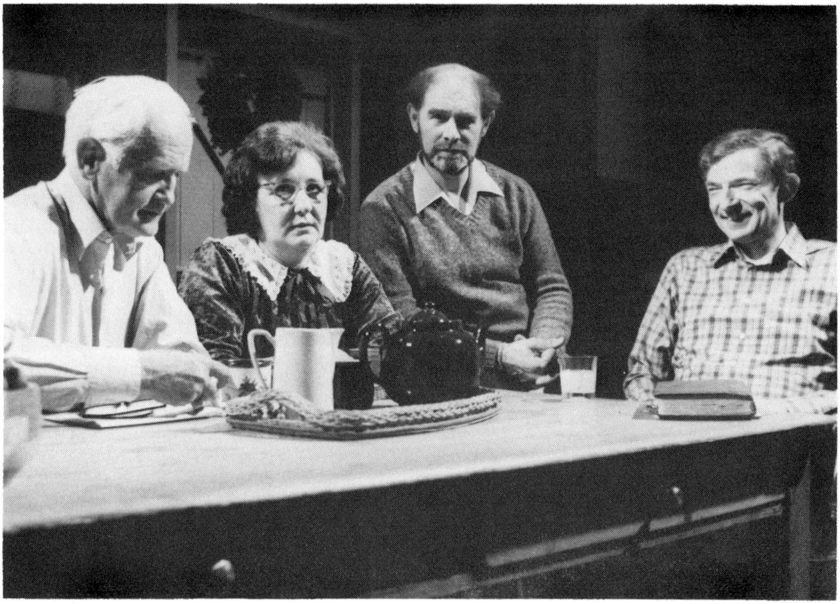

*David Watson joined the cast of 'Company' (TVS end of broadcasting 'epilogue') to make a series with George Reindorp (left) and other guests. The programmes have been put together in* The Best is Yet to Be *(TVS)*

6. *Method of presentation will vary according to the size of the group.* At present television/video presentation to large audiences presents difficulties. Even the largest television receiver is usually too small for effective display. Captions very often cannot be read; fine detail is lost. The large screens, presently available are not altogether satisfactory. We are assured, however, that the day of the very large high quality flat screen is not far away and that we may expect to see these mounted on the walls of halls, etc. However, given the advanced technology, we may still ask, is this what we want for training purposes? I can see churches making much of the large screen for the purposes of evangelism, worship and the screening of major feature programmes. However, I find the optimum size of

adult learning groups to be around 24–30 people. Usually, in conferences of 60–70 people, I timetable for duplicate sessions in specialised disciplines. With 24 people a 26 inch television screen is perfectly acceptable, particularly if elevated. As a personal preference I work with a flip chart or whiteboard alongside the television/video trolley and like to note down and display comment, query and reaction to any video piece of a substantial nature. Conversation is possible really only in these smaller groups. Larger groups, even in plenary session, tend to become audiences (or even congregations!). If the leader is adopting an enabling style then smaller groups are better. If he is working with larger groups then the entire presentation will tend to be more formal, more directive and, I believe, his use of video more stilted, almost to the point of being a film-show – and *that* is to alter the experience drastically.

7. Some video programmes are highly didactic, adopting a problem-solving approach. This invites the possibility of *placing the group in competition with the video.* For example, in *Choosing the Way*, I set out to look at religion and the Church from the point of view of young people in their mid teens. What kind of views did they hold of the Church, of God, of the world; what were their questions? Beginning from this perspective, the video poses problems. In working with youth leaders it is interesting to switch off the video at that point and invite the group to respond to the challenges expressed by the young people. Only then, you show the remainder of the unit and compare the 'answers' afforded on the programme with those put forward in the group discussion. Although I have not yet done so myself, it would be most interesting to develop a series of bible studies along these lines – if you like, a highly visual form of Hans Reudi Weber's *Experiments in Bible Study.*

8. There is at least one other use of video in training which deserves a mention and that is *the home-made video.* Here there is no great concern for quality, professional direction and editing but rather the recording of an exercise with a view to learning from the play-back and to formulating basic principles for the more successful performance of the exercise. This may be used to study simulated experiences such as parish visitation, after the manner of the professionally produced *So, You're Going to be a Visitor*, or conflict situations as a means of deepening relationships, as in *Fighting Talk.* A home-made, fly on the wall style, documentary of a typical Sunday school morning would probably stun the teaching staff!

## In conclusion

In a world drawn toward specialisation and sophistication in communication skills the Church has a duty and a need to keep abreast of the most recent developments. This chapter has sought to focus on the use of television/video as an educational resource. We have been concerned more with purposive education as such rather than the educative nature of television generally. In particular we have examined video resources as a tool for the training of Christians offering themselves for particular areas of service, whether full-time or voluntary spare-time. We have considered video as one important possibility in the 'resource battery' always recognising that, ultimately, the educator himself is the key resource.

The last word must be concerning the future. Without question visual learning will develop very substantially. It is inevitable that even the most hesitant of Christian groups will be drawn to use these new methods of communication and teaching. I would simply make the plea for greater understanding of the multi-faceted potential of this medium and in particular for the development of productions that will offer to Christian educators, and so to congregations on the receiving end, resources of the quality, variety and perceptiveness that the work of Christian nurture deserves.

## Some further reading

Video in the Churches: A Survey, *Brian Brown* (Oxford Polytechnic Television Research Unit and Trinity Trust).

Television Studies in Scottish Schools, *ed. J. F. Murray*, Occasional Working Paper No.6. (Scottish Council for Educational Technology).

Television: Technology and Cultural Form, *Raymond Williams* (Fontana, Collins).

What's this Channel Four? (an alternative report) *ed. Simon Blanchard and David Morley* (Cornedia Publishing Group).

Experiments in Bible Study, *Hans-Reudi Weber* (W.C.C.).

## Videos referred to in chapter

| | | duration | distributor |
|---|---|---|---|
| Help Me to Teach! | | 1:05 | Ch Scot Education |
| Meet Cliff Richard | | 0:25 | Ch Scot Video |
| The Very Gate of Heaven | | 0:30 | Ch Scot Video |
| Meet Dr Allan Boesak | | 0:30 | Ch Scot Video |
| Choosing the Way | | 0:48 | Ch Scot Education |
| Jesus Then and Now | six tapes, each | 1:00 | Trinity Video |
| So You're Going to be a Visitor | | 0:22 | Ch Scot Video |
| Fighting Talk (on tape: People Need People) | | 0:13 | Ch Scot Video |

## Some sources of video in training for Christian service

Few suppliers/producers of Christian video cater for the needs of Christian education and training other than in the most general and 'educative' way. On the other hand some secular organisations produce useful programmes which may be used to great effect.

You may wish to write to the following organisations for details and catalogues of training/education programmes currently available.

See 'Programmes for Education and Training' from BBC Enterprises (but be clear about costs in advance; BBC videos tend to be expensive). Categories of particular interest include Child Development, Religion, Sex Education, Social Issues, Astronomy and Third World. In the *Schools and Education* section examples of useful tapes are 'A Lesson for Teacher', 'If at Last You Don't Succeed, You Don't Succeed', 'Living in a Developing Country', and 'Have a Drink?'

| | | |
|---|---|---|
| BBC Enterprises Ltd | Woodlands<br>80 Wood Lane<br>London W12 0TT | 01-743 5588 |
| Church of Scotland Video<br>Church of Scotland Education | 121 George Street<br>Edinburgh EH2 4YN | 031-225 5722 |
| Scripture Union | 130 City Road<br>London EC1V 2NJ | 01-250 1966 |
| Trinity Video | Westbrooke House<br>76 High Street<br>Alton<br>Hants GU34 1EN | 0420 89141 |

5

# Working with children and teenagers

Anton Baumohl (Scripture Union)

Some years ago the television set was described as 'a latter-day nanny' in front of which parents dumped their children as they got on with life. In recent years this nanny has become available to the church in the form of video. But in an age when we may try any means to capture the attention and interest of restless children and young people, let us not use the small screen as an easy means of solving difficult problems.

Children and teenagers are, however, avid television viewers and have grown up with the small screen as an instant companion. This makes video a potential ally when used to help young people understand the value and purpose of the Christian faith. It enables the most spectacular productions to be used in church or home where the viewing is for entertainment but where the message coming through is essentially Christian. It enables small age-based groups in club or Sunday school to open their minds to important areas of Christian teaching in ways that will enable them to understand the implications for their own lives.

In Sunday school, holiday clubs or midweek activities the video proves a winner because of its ease of handling. There are no filmstrips or tapes to thread, no lamps to blow, no blackouts to erect, no multiple controls to manipulate with limited hands. The video cassette is easy to insert and the recorder simple to operate. Sound and visuals are on one tape and only one recorder plus a television set are required. The major limitation in these situations is that the video is still, essentially, a small group machine, the actual size of the group depending on the size of the television screen used.

It is possible to get 25 youngsters around a 22 inch screen but it is not suitable for a family service gathering of 50 or more. When numbers get that large you have to invest in more sets. Large screen video for medium and large audiences does exist and its advantages and disadvantages have been discussed in other chapters. Throughout this chapter we shall be assessing the use of video in small groups of up to about 30 but as you bear that in mind don't rule out the possibility of larger audiences.

The recorded cassettes can usually be bought or hired. Most producers sell their cassettes directly to the public or through outlets like Christian bookshops. The majority of programmes can also be hired, either directly or through video libraries which operate a mail order service or are in outlets like bookshops, church resource centres, etc. It is obviously cheaper to hire especially if you will only be showing the video once.

## Preparing for a video showing

General guidelines have been given elsewhere. Some points that need to be borne in mind when showing video to young people include:

1. Before the meeting run the cassette through (in part if it is too long to view the whole thing) and check you have the correct cassette. Also check the *tracking*; in effect this is the way the video recorder picks up the recording from the cassette and the recorder may need adjusting to align this accurately. One sign that this is necessary is when there appears to be constant interference as the film is showing. Children can be particularly intolerant of a substandard quality. Find the tracking adjustment knob on the recorder and tune as necessary.

2. Place the television set high enough so that everyone in the group can see it, without having to peer around others' heads. If it is possible to seat everyone on the front row all the better. Make sure the set isn't so high that viewers get neck-ache looking up!

3. Ensure that the young people's viewing won't be distracted by people walking behind the set, so avoid putting it in front of a window, door way or popular route between two rooms. Although it isn't necessary to black out the room, if practical blackouts will help prevent distractions then use them.

4. Place the recorder near the television set but out of reach of prying fingers and test for volume level. Remember to increase the volume a little if testing in an empty room, as a room filled with bodies absorbs some of the sound sometimes making an adjusted recorder inaudible.

Two other points have not been mentioned elsewhere that may seem obvious or trivial but do contribute to a successful viewing both for yourself and for others after you. Before the meeting check that you have a cassette that is compatible with the format of the recorder. After the session rewind the cassette before returning it to its box.

## Using video in teaching

Some may rightly argue that children and teenagers spend long enough in front of a television all day so why extend that through use in Sunday school or bible class or club. In using video for teaching we *do* need to question our motives. Is this video programme really going to teach our children or teenagers something they need at the present or are we just using it to save us time and preparation? Is this one amongst a variety of teaching methods I use or have I got stuck in a rut?

Video programmes are best used when (a) they are well produced ones that fit the teaching programme that we are following and that has been designed to suit the needs of *our* youngsters, (b) they bring variety to our methods of teaching, so helping to retain the interests and enjoyment of the youngsters with whom we work, (c) when the occasion is special and we want to mark it with a special presentation, e.g. an anniversary, the culmination of a holiday club or a special weekend's activity.

In teaching sessions the video will only be part of the programme never the whole, and the best videos for teaching are short – up to about 20 minutes. This allows time for viewers to react to the material they have seen. Remember children and teenagers need a high level of participation if they are to learn effectively. Video, however, is a passive medium and must be followed up by more participatory methods.

## Teaching children

There are still relatively few good video programmes for children. Among the most popular are cartoon series which may be either animated or compiled from still pictures. In the first category are the *Superbook* series from the United States. These were originally a broadcast television series and use animated cartoon figures to tell the stories of the bible from the viewpoint of modern children who find an old copy of 'superbook' and in reading it get drawn into the actual stories. The result is a colourful, lively, and original approach to recounting bible stories. Each cassette has three 40 minute stories which are rather long for a teaching session but worth spreading over two sessions if necessary.

*Luke Street* is a series of five short filmstrips very effectively transferred to video and which lend themselves to lessons or epilogues. Each story is taken from Luke's gospel and is introduced 'live' by Roy Castle who also narrates the stories. The five stories are self contained, each lasting seven or eight minutes. A booklet with suggestions on how to use the video in a teaching programme is included with the video. On the same Scripture Union cassette there are also four short stories from Matthew's gospel under the series title of *Follow the Leader* and seven stories from John's gospel called collectively *Signposts*. That's sixteen stories in total on one cassette!

These stories can be used with children from five or six upwards but are ideal for 7–11 year olds. In order to get the best out of these videos they need to be followed up and the teaching reinforced and applied. Below is a typical 40 minute programme that might be used in Sunday school or on a holiday club:

| | *minutes* |
|---|---|
| Opening worship (song, prayer). | 3 |
| Video of one story. | 8 |
| Song/chorus. | 1 |
| Activity based on story – model making, colouring, drama, quiz, collage, puppets. | 20 |
| Short illustrated talk. | 5 |
| Closing worship. | 3 |

After staring at a television screen for even a short while children get totally absorbed and need a little time to 'focus' back on the 'real world' – hence the chorus directly after the video showing.

The activity period provides participation for reinforcing the story, and with craft-based activities it allows the teacher to walk around talking informally to the children about what they have seen. The short illustrated talk provides an opportunity for applying the story to today, although this is often best integrated into the activity work, e.g. through drama, etc.

Where time is short such as at the end of a club evening the learning value of the video would be enhanced with some quick-fire questions to refresh memories – "Who said . . . ?" "What did . . . ?" "Why was . . . ?" "What happened when . . . ?" "How would you . . . ?"

In looking for good teaching material for use with children (and teenagers) don't neglect the secular agencies. Many cartoons and stories shown as children's television have good Christian moral teaching to offer. Look around the video hire shops. If you see a suitable item on the television, contact the broadcasting company involved and ask if it is available on video cassette. *The Selfish Giant* is an Oscar Wilde children's story that has been shown on the television and has good teaching potential. Follow-up an allegory like this by looking at a bible story that illustrates the same teaching or some words of Jesus or Paul that encourage the same values or behaviour and make a clear link between the story and the Bible's teaching. The link doesn't have to come in the form of a little homily but could again be a piece of drama or craftwork that takes the child from the world of the cartoon into the world of the Bible. Other programmes from secular distributors are mentioned in the final chapter, and there is a comprehensive list in the *Guide to Christian Video.*

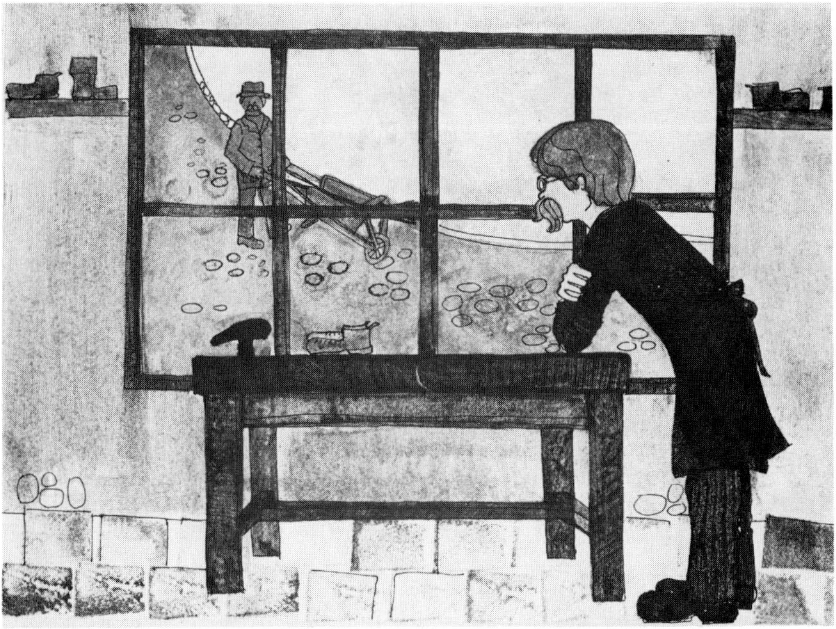

*The shoemaker Papa Panov, alone on Christmas eve, hears the voice of a visitor in* Papa Panov *(Scripture Union)*

Although children will sit absorbed by the television screen for hours they do not retain vast amounts of detail. So if a video is being used as part of a teaching/learning programme I would, again, stress the need to keep the video section relatively short and followed up by other activities. If the teaching required is part of a lengthy video recording either use only the essential section, or divide the showing up into parts interspersing each part with other activities.

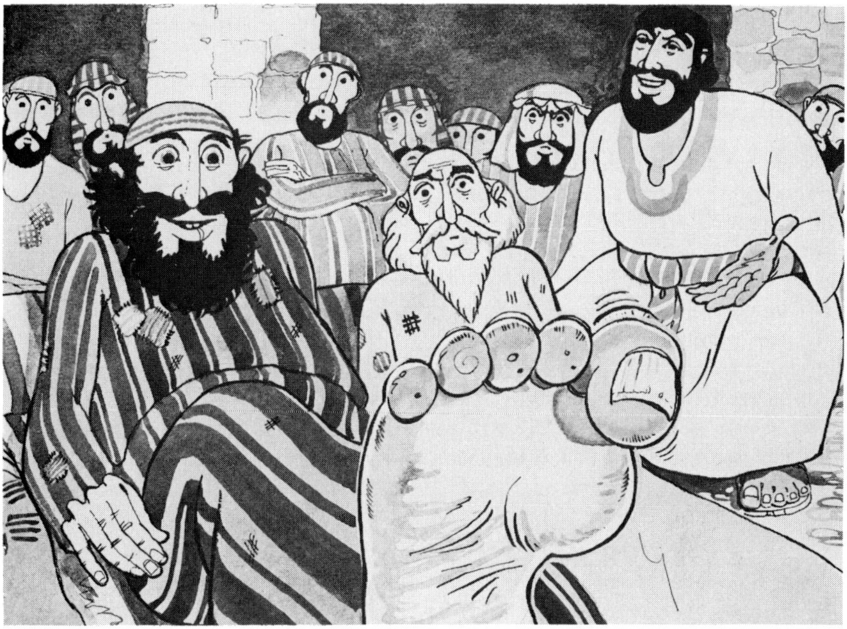

*The seven signs or miracles that St John picks out from Jesus' ministry are the basis of* Signposts *(Scripture Union)*

## Teaching teenagers

A greater variety of video material exists for use with teenagers than for children and many programmes produced for adult groups are equally suitable for the bible class, confirmation or church membership group, informal discussion groups or youth club epilogue. Cartoons still form a valuable means of communicating. CPAS' *Solly Silverberg's Old Time Stories* and *Hipperty Dog* stories, and Scripture Union's *The Challenger*, *The Champion* and *Number One* are all filmstrips that have been very successfully transferred to video and all form excellent discussion starters. Some of these present biblical teaching through allegory or in parable form while others retell a bible story through cartoon characterisation with lacings of humour. Videos like *The Champion* are backed-up with songs rather than through spoken script.

A series of five short animated cartoons under the title of *Screen Test* are produced by CTVC, excellently conveying thoughts through a combination of simple animation and music. They are all on one 20 minute cassette.

Drama on video also makes excellent teaching material for teenagers. Examples here are the sketches of *Riding Lights* (Lella) and the mime and music of Geoffrey Stevenson and Graham Kendrick on *Seven Ages of a Fool* (Scripture Union). These video programmes present 'live' drama with professional drama groups providing a high level of performance. The Christian message is often presented through modern day situations and with the use of modern parables. The few Fact and Faith films available on video provide some interesting links between science and Christianity and the photography is usually quite spectacular (especially in *Journey to the Stars* and *A Mystery Story*). Background teaching can be stimulatingly supplied through documentary recordings of the holy land. Then there are a growing number of short personal testimony videos featuring real-live Christians sharing something of how they came to faith. (Keep an eye open for short rather than long personal stories.) A useful example for sports-conscious groups is *Christians in Sport*.

Whatever type of video you choose look upon it as a springboard for discussion or, if the group is amenable, for dramatic presentations of their own. Again the video presentation needs to be relatively short in order to allow time for other items in the programme. If following a video with discussion, the discussion may be left open and general, such as "What did you think of that?" "How do you react to that ?" or can be structured guiding the group through different areas but still being prepared for a variety of answers.

In structuring a discussion the following phases may prove useful.

(a) focussing on the video recording itself:
"What do you remember about . . . ?"
"How did you feel when . . . ?"
"What major theme would you say came through?"
"How would you sum up the video and what it was saying in one sentence?"

(b) focussing on the experiences of those watching:
"Have you ever had that sort of experience?"
"Could you identify with anyone in the story?"
"How would you have have dealt with . . . ?"

(c) helping people apply it to their own lives:
"What did it teach you/say to you?"
"How will that video influence you during this week?"
"If we/you were to take it seriously how would we/you have to change?"

If it proves difficult getting a whole group to enter into discussion then get them to work at questions in pairs or small groups and then to report back to the whole group.

When using drama to follow up a video a number of ideas could be developed. The most valuable follow-up is drama that takes little

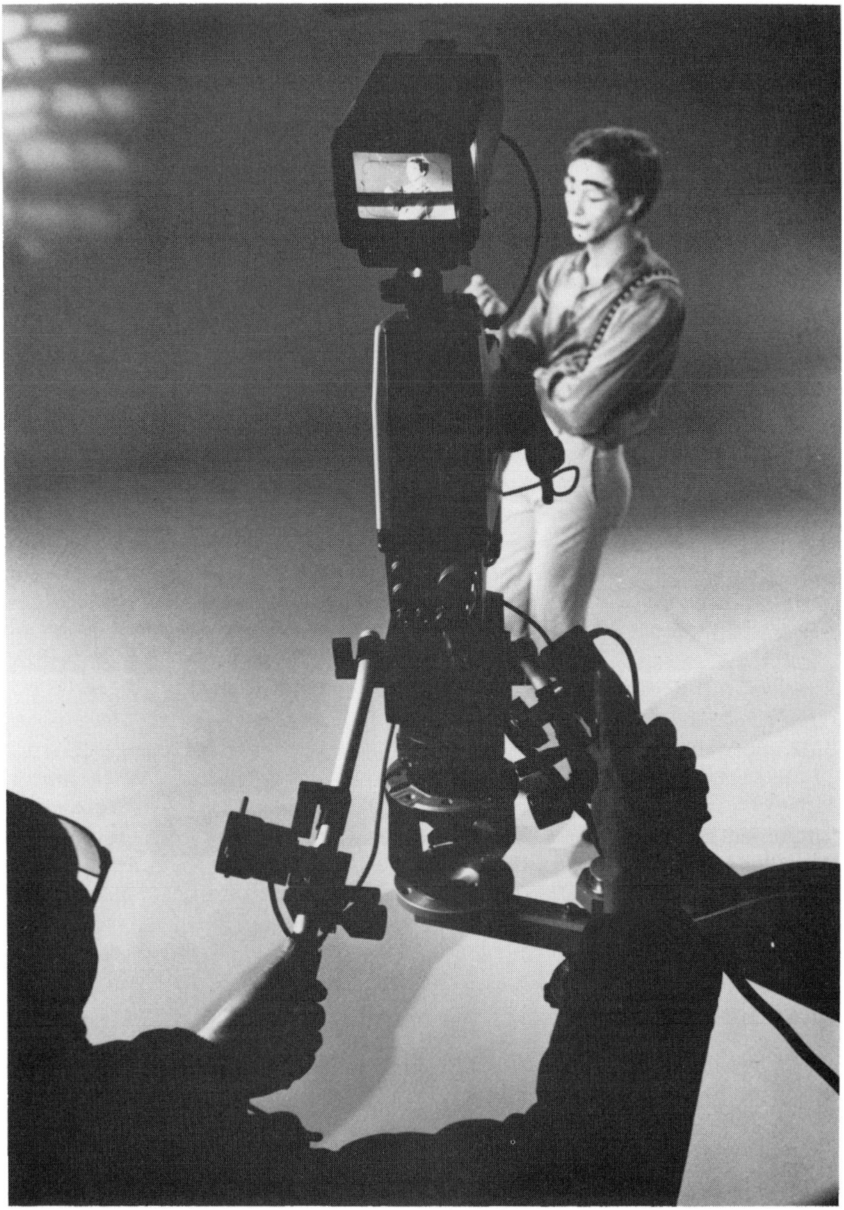

*Geoffrey Stevenson, the mime artist, as seen by the cameraman in the making of* Seven Ages of a Fool *(Scripture Union)*

preparation, few props and isn't for performance to some audience – its value is to those who take part in it. There are various effective ways in which a group can participate.

1. They can prepare a mime that communicates the key message of the video.

2. They can prepare a mime that retells the story seen on video (younger teens).

3. They can prepare a playlet (with dialogue) that explains the main message of the video.

4. They can prepare and act a parable that provides the same teaching but in a different context.

The drama may in turn be followed-up by discussion.

## Video entertainment

In our concern to ensure we teach the Christian faith effectively and with urgency we should not neglect the place of entertainment in the life of the church. Entertainment, leisure and fun are valuable in their own right and are part of the recreational process the church may need to encourage actively. Video programmes have been one of the biggest growth areas in the leisure industry over the past few years and although the Christian world has not met this challenge by providing good 'Christian entertainment' there are video programmes available that are as good as a night at the cinema.

For Christians it is often the setting or context that determines whether a video is being used for entertainment or teaching. In a structured teaching time, the aim of the video is to stimulate thinking, promote discussion and encourage learning. Where Christians come together to relax, have a social evening or enjoy time with their families the use of video may have no other function than to provide an opportunity to relax. In practice it is impossible to say that people will not learn from an evening's video entertainment. At the very least the video can communicate Christian values or offer a contrast so that the difference between Christian and non-Christian values become clearer. These videos may also have the effect of convincing people of the reality of the Christian faith and so be evangelistic or help to lead Christians on in their understanding and so be discipleship tools.

Videos can be used for sheer entertainment in almost any setting – a children's party or holiday club, part of a wet weather programme, teenage youth club, or when the generations are together at a church houseparty, harvest supper, indoor picnic, church weekend, Christmas party or in the home itself.

A few very good full-length films exist on video for use either with children on their own or in mixed generation groups. The two most notable are *Tanglewood's Secret* and *Treasures of the Snow* which are videos taken from original films lasting about 1½ hours and produced to professional quality. Both are adventure stories involving children and contain clear and naturally incorporated Christian teaching. They could be shown to a totally Christian or mixed audience with confidence.

Many of the cartoon based videos mentioned earlier can also be shown to children without further educational back-up although they are often devised with a teaching programme in mind. For the Christmas party try Scripture Union's *Papa Panov,* a video of a filmstrip.

Teenagers again have a wider choice than children with excellent full-length films produced by Christian groups with a clear message. Amongst the most popular are *Joni,* the compelling story of Joni Eareckson and her fight to cope with disablement caused by a swimming accident, *The Hiding Place,* the story of Corrie Ten Boom and her life under the Nazis during the second world war, and *The Cross and the Switchblade* which tells of work in the gang and drug infested parts of New York.

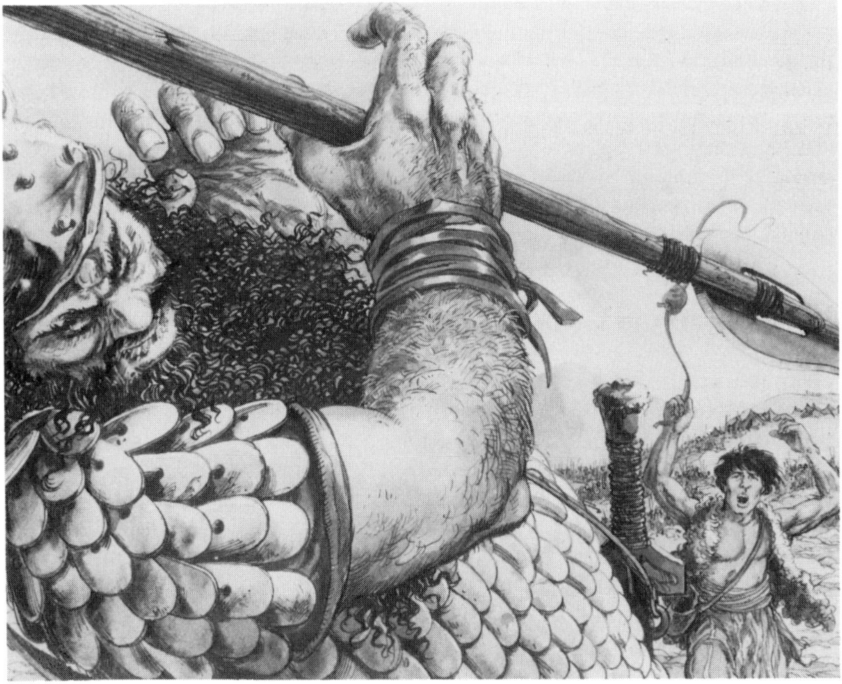

*David's stone hits home in* God's Story *(Yorkshire Television)*

There are also a number of films produced for the big screen by recognised secular feature film-makers that form ideal focal points for entertainment evenings. In recent years the best known of these is *Chariots of Fire* which tells of the story of Olympic runner Eric Lidell who struggled to relate his Christian faith to a love of athletics.

A great deal of teenage leisure time is taken up with music, and the pop world itself has made big strides in linking contemporary music with video film to the extent that some pop music is now dominated by its visual rather than audio dimension (reflected in the rather cynical phrase, 'nice video, shame about the song'). A few video/music links exist in the Christian video world and include: *Gospel Road* featuring the music of Johnny Cash, *Jesus Christ Superstar* the rock musical of the original stage show about the life of Christ, *Greenbelt Live* that highlights some of the musical items at the Greenbelt Christian pop festivals and *Seven Ages of a Fool.*

If some of the longer videos won't fit into the time available during the normal meeting time you might consider serialising the film, showing 20 or 30 minute sections for two or three weeks. This has been done successfully with *The Cross and the Switchblade* and *Jesus of Nazareth.*

It must be stressed, yet again, that video is essentially a small group medium unless a number of television sets or large screen video projection is used. This may limit the use of video for large scale church entertainments. In the children's or teenage club this can be overcome by making video one of a number of options. So that while some are playing games, drinking coffee, doing craft work, a video programme is shown in a side room for those wanting to sit and watch. The most obvious small group for video consumption is the family unit.

## Viewing in the home

Homes form natural small intergenerational units in which video may have a place. With so many complaints about the standard of programmes on the television networks, Christian families may prefer to have a Christian video in order to 'raise the moral tone' of the family's viewing. Parents with children may like to show videos that they can talk about afterwards. A video of bible stories at bedtime may not (hopefully) replace reading a story to the children, but it may be an interesting variation to use on occasions.

One problem that parents may have is actually obtaining the videos to show, as most are not available through High Street shops. Some very good comprehensive Christian video libraries exist and local Christian bookshops may also offer a video loan service. These should be encouraged and supported wherever possible, but it may be more forward thinking for a church or even a group of churches in a locality to start their own video resource library and buy tapes from distributors and video wholesalers. It can then hire out videos to youth and children's groups and to families, with the possibility of getting Christian video into non-Christian hands.

## Producing your own videos

If you really want to involve and excite your children and teenagers with video one sure way is to get them involved in producing the programmes themselves. This is the ultimate in participation. You will need some basic equipment which if it is not to hand can probably be hired from a video shop, a resource centre or the local education committee youth service office.

Here's what you will need:

1. Domestic video camera.
2. Portable recorder.
3. Camera tripod.
4. Lights (two ciné lights of about 1000 watts each) if you want to video indoors.
5. Blank video tape.

Filming with video has the big advantage that the film requires no processing and can be viewed immediately after 'shooting'. Cameras can usually be handled by junior and teenage children after a little practice.

With professional video the film has to be edited to ensure proper sequencing of shots. If you keep your videos short and don't mind the odd mistakes editing doesn't need to bother you at all. If, however, you want to do some editing using two video recorders, play the filmed video on one recorder and record the pieces you want, in the required sequence, on the second recorder.

The real value of using video in this way is in the doing and in viewing and discussing the finished product. The actual quality of the finished product is of secondary importance.

Here are some programme ideas:

1. Video a puppet show using a bible or modern story. Spend a weekend or series of evenings running an intergenerational activity. Divide the group of adults/teens/children into three groups. Group one makes simple puppets from cardboard tubes, fabric, paint, etc. Group two produces scenery and props. Group three works on the script. When all is ready re-divide into two groups. The first group manipulates the puppets and the second takes turns in using the

video camera. Show the results (after editing if necessary) to those involved in production as well as to younger children who couldn't take part in the production.

2. Ask members of the teenage group to video a typical church service while others join the congregation for worship. Replay the film at a youth meeting as a discussion starter on worship and the church. Use discussion questions like: "What did it feel like taking part in that service?" "What do you understand by the word worship?" "How could the service have been improved to encourage more worship?"

3. Ask a group of juniors to mime a bible story and show them the results. Use the video as part of a lesson on some aspect of the Christian faith.

4. Ask a group of teenagers to role play a common relationship problem, e.g. teenagers arguing with parents over what time they should be home. Others video the role play and play it back as a stimulus for discussion in groups.

5. Take the video camera into the streets and interview people about their beliefs and attitudes on a particular topic (Christmas, easter, church, young people, Jesus, etc.).

In our work with young people and their families we should be concerned to use all means to explore and explain the Christian faith. Video is a contemporary medium and as such provides a valuable instrument for Christians in their work of teaching and evangelism.

## Videos for children

| | duration | distributor |
|---|---|---|
| Superbook 1: *The Beginning*: How it all began; My Brother's Keeper; The Flood | 1:20 | Bagster |
| Superbook 2: *The Patriarchs*: The Test; Here Comes the Bride; Double Trouble | 1:20 | Bagster |
| Luke Street; Follow the Leader; Signposts | 1:23 | SU |
| The Selfish Giant | 0:27 | Videoform |
| Duke Starduster and the Margins of Nisan | 0:36 | Falcon |
| Duke Starduster and S Factor Positive | 0:36 | Falcon |
| God's Story (old testament) | 2:55 | Yorkshire TV |
| God's Story (new testament) | 1:40 | Guild |
| The Vision | 1:25 | Church Army |
| Tanglewood's Secret | 1:20 | Bagster |
| Treasures of the Snow | 1:55 | Bagster |
| Papa Panov | 0:12 | SU |

## Videos for teenagers

| | | |
|---|---|---|
| Solly Silverberg's Old Time Stories | 0:34 | Falcon |
| Hipperty Dog 1 | 0:36 | Falcon |
| Hipperty Dog 2 | 0:34 | Falcon |
| The Challenger; The Champion | 0:45 | SU |
| Number One; The Stranger; In the Bin | 0:41 | SU |
| Screen Test | 0:23 | CTVC |
| Riding Lights | 1:15 | Lella |
| Seven Ages of a Fool | 0:37 | SU |
| Journey to the Stars | 0:10 | FFF |
| A Mystery Story | 0:10 | FFF |
| Christians in Sport | 0:30 | Chr. in Sport |
| Joni | 1:52 | Internat. Films |
| The Hiding Place | 2:05 | Internat. Films |
| The Cross and the Switchblade | 1:45 | Gateway |
| Chariots of Fire | 1:58 | CBS/Fox |
| Gospel Road | | Bagster |
| Jesus Christ Superstar | 1:47 | CIC |
| Greenbelt Live | 0:50 | Grenville |

## Some distributors

| | | |
|---|---|---|
| Bagster Video | Westbrooke House<br>76 High Street<br>Alton<br>Hants GU34 1EN | 0420 89141 |
| CBS/Fox | Perivale Industrial Park<br>Horsenden Lane South<br>Greenford<br>Middx UB6 7RU | 01-997 2552 |
| Christians in Sport | PO Box 93<br>Oxford OX1 1QX | 0865 240407 |
| Church Army | Independents Road<br>Blackheath<br>London SE3 9LG | 01-318 1226 |
| CIC Video | UIP House<br>45 Beadon Road<br>Hammersmith<br>London W6 0EG | 01-741 9333 |
| CTVC | Beeson's Yard<br>Bury Lane<br>Rickmansworth<br>Herts WD3 1DS | 0923 777933 |
| Fact and Faith Films | 120 The Rock<br>Bury<br>Lancs BL9 0PJ | 061 764 1538 |
| Falcon | CPAS Mission at Home<br>Falcon Court<br>32 Fleet Street<br>London EC4Y 1DB | 01-353 0751 |

Gateway Films are distributed by Bagster Video
Grenville Films are distributed by Bagster Video

| | | |
|---|---|---|
| Guild Home Video | Guild House<br>Oundle Road<br>Peterborough PE2 9PZ | 0733 63122 |
| International Films | 235 Shaftesbury Avenue<br>London WC2H 8EL | 01-836 2254 |
| Lella productions are distributed by Trinity Video | Westbrooke House<br>76 High Street<br>Alton<br>Hants GU34 1EN | 0420 89141 |
| Scripture Union | 130 City Road<br>London EC1V 2NJ | 01-250 1966 |
| Videoform | Unit 4<br>Brunswick Industrial<br>Park<br>New Southgate<br>London N11 | 01-368 3318 |
| Yorkshire Television | The Television Centre<br>Leeds LS3 1JS | 0532 438283 |

6

# Video in evangelism

Stephen Maxted (Spring Harvest)

The themes of the previous chapters have all been related in some way to the idea of our growth as Christians and equipping and preparing us for living and serving. From this we should be looking now to see how we can become more effective witnesses. It is apparent that video has an important part to play in this at all levels of communication, from one-to-one to making the speaker at the mass rally more accessible.

Evangelism involves me as an individual being a witness to the saving power of Christ, and therefore video, or any other medium, must not get in the way of personal communication. However, just as printed material has been used in the past as an aid at all levels of communication, so now video can be used as a most effective tool.

## One-to-one

This is where I believe evangelism begins. It begins with me asking God for an ability to share Jesus with others; an opportunity to get to know and make contact with people who don't know him and for opportunities to share the love of Christ and lead them to a point where they can accept God's free gift of salvation for themselves.

As friendships and conversations develop you may find yourself in a situation where you might feel inadequate in giving the sort of explanation that you would want to give on say "why you believe

Jesus Christ rose from the dead''. Your friend may not necessarily be a person who reads books and certainly not anything that might appear to be a weighty theological tome. This is where a video such as Eric Delve's *Why I Believe Resurrection is True* produced by Spring Harvest could be a terrific help.

In this situation you can either lend your friend a video tape that you have borrowed from a library or say something like ''I'm finding answering that question difficult myself but if you would like to come round to my home tomorrow you can see a video that explains it much better than I could and answers all the questions you have been asking me''. Videos don't solve all problems but because of the wide range of apologetic and teaching material that is available they can be used as a tool to answer questions that are put to you in personal conversation.

Maybe your friendship developed along different lines and you have invited your neighbours in for an informal coffee evening or dinner party and in these sorts of situations questions casually thrown out like ''would you like to see . . .'' with a brief explanation of the content of the particular programme would be a good introduction to help you explain your faith.

Perhaps the relationship with a neighbour is more tenuous. Try offering them a video cassette to watch in their own time without you being present and you may be surprised how readily they will accept it from you. Many will take a video of a feature film such as *The Hiding Place*, watch it in the privacy of their own home and then when giving it back to you will quite openly discuss the contents. You can be sure of an interesting conversation when discussing the following films with them: *Joni, The Story of Jesus, The Cross and the Switchblade, Born Again*, or *Chariots of Fire.*

## Group events

In planning outreach we must obviously move from our personal contacts and people that we are talking to at a one-to-one level through to events that might be organised by a ladies' group, men's forum, housegroup, young people's group, Christian Union, etc. Such events will vary tremendously in nature. Examples might be a party, an evangelistic bible study, a staggered supper where each course is eaten in a different home, a coffee party which can be held in a morning, afternoon or evening, a Christian Union meeting in a school, Know Your Neighbours night, a street party, and obviously many more.

Video can play a useful part in the presentation of the Christian faith in all of these kinds of events provided that the material is chosen very carefully and the introduction and follow up is well handled. The person choosing the programme must first of all ask the question "is it suitable for the people that I am going to invite to my meeting?" To put it at a very basic level, one wouldn't use a stretching apologetics lecture to show to five year old children. You may feel that is obvious, but I have been surprised at how little thought has sometimes gone into the prior vetting of material.

Running time is another very important factor since a video which runs too long can actually kill the meeting even if the content of it is very good. Experiences differ, and again it is a matter of knowing the people who will be present. Personally I would ensure that any video presentation that I was using for a party would be short – probably no more than 20 minutes long. For these sorts of meetings I still feel that the video version of Scripture Union's *Number One* would take a lot of beating. This 17 minute cartoon narrated by Derek Nimmo is intended for teenagers and adults. An alternative would be *In the Bin*, on the same cassette and for a similar audience, in which an ordinary office worker under the very believable name of Arthur Stanley Grimble meets a dustman who is to strip away the mask that he wears.

*Video suppers*

Videos have been widely used at the Dalneigh and Bona Church of Scotland in connection with outreach work that culminated in a week of mission. In order to encourage people to meet together the Rev Fergus A Robertson organised a number of video suppers.

At a preliminary briefing meeting with the people who were to be the leaders, or hosts, at these suppers, the minister shared his ideas and experiences. Each supper was to be held in an individual home and neighbours and friends would be invited for the evening.

The evening would begin at 6.45 when an informal supper would be served. This took the form of soup and roll and some kind of sweet. To encourage the individuals to mix they had to go to the kitchen rather than be served where they were sitting. So the evening began with a time of mingling and sharing, barriers hopefully being broken down over supper. About three-quarters of an hour was allowed for this.

Then the group settled down to watch a film on the television set. It was suggested that the film should not last longer than two hours, although in some groups longer films were shown successfully. A wide variety of films were used and the leaders of the various groups were encouraged to select from the video library those programmes that they thought would be most suitable for their particular group.

Some chose to show an interview, some an actual feature film. Others chose a testimony and some even cartoons.

After the film, coffee or tea was again served, once more from a central point so that people had to get up and go for it. Again this provided an opportunity for mixing and talking. It was at this point that conversations hopefully developed. Sometimes there was a period of questions immediately after the video, but this was not unduly prolonged. As soon as there was any sign of its drying up, then coffee would be announced.

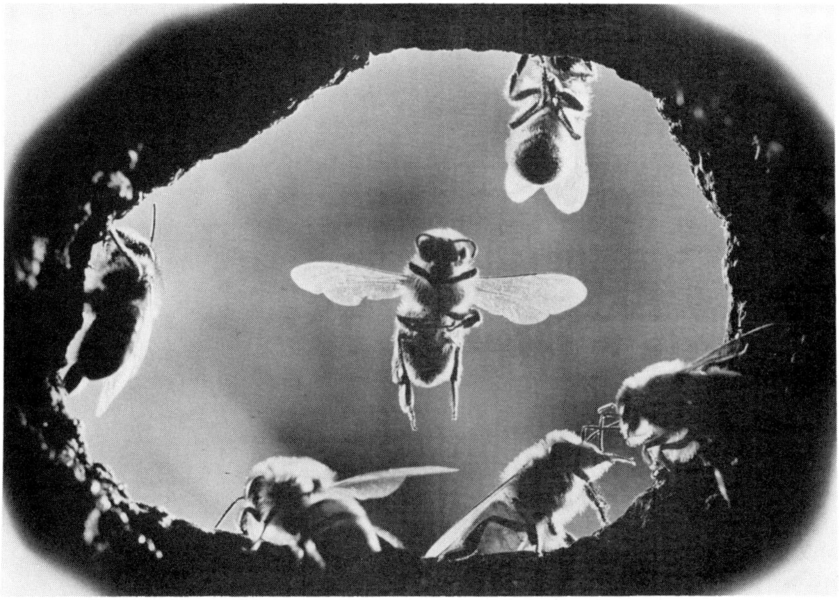

*While mankind struggles with the problems of human behaviour, bees have a developed social structure that works.* City of the Bees *is a Sermon from Science from Fact and Faith Films (Moody Institute of Science)*

The leaders were encouraged to hold at least two of these video suppers during the months leading up to the mission. The video provided a safe focus for the evening during which people could relax and not feel threatened. By the time that two such suppers had been held in a group, the members of that group were beginning to know one another and feel relaxed in each other's company. Consequently during the mission itself it was possible to hold, not video suppers in the home, but actual meetings with members of the mission team sharing in the fellowship and speaking to the people.

The role of the video then was very much to break down barriers, to attract folk to a fairly low key and non-threatening evening and to introduce them to the warmth of Christian fellowship.

This would be an inappropriate place to undertake an in-depth examination of how to lead a discussion or other small group, but in organisation of any group meeting at which a video is to be shown the following points should be considered. Some have been made in the chapter on housegroups, but their repetition may serve to underline their importance.

1. Preview the tape before the meeting so that you become familiar with its content. If you are taken by surprise by the tape then there is no way in which you can pray towards or plan the discussion that is to follow.

2. Before the meeting starts, place yourself unobtrusively in a position where you can observe the reaction of other people.

3. Watch out for clues that tell you that your audience is getting bored.

4. Have pre-planned questions which can be asked in the middle of the programme if you feel that attention has lapsed so much that the programme needs to be stopped.

5. Watch out for positive reactions that might be used in later discussion.

6. At the end of the video some of the following questions might be useful to start a discussion:

How was the film relevant to our own experience?
What biases did it show?
How could it have been better balanced?
What particular aspects of the characters made them believable?
In what ways did the film stimulate you?

Note how all these questions are phrased to require an answer more than "yes" or "no" and to encourage a positive contribution to a discussion.

7. You might possibly ask people what they felt was the main message of the film. Be sensitive, never push too hard for an answer but on the other hand a good discussion leader should ensure that nobody sits in the corner thinking that their contribution is not wanted.

In the earlier chapter on video for housegroups, some suggestions were given of programmes that might be used with an audience that consists of a majority of Christians. Unfortunately, there are very few purpose-made evangelistic programmes, other than feature films that have been transferred into video format or recordings of evangelistic rallies. Among these few are two by Luis Palau that consist of sets of short talks – *Video Evangelism* and *An A.B.C. of Evangelism*, and one by Peter Anderson, also containing a number of short talks – *The Relevance of Christianity Today*.

## Church events

Moving on from small group meetings, we need to take a quick look at the way in which video might be used in meetings for the whole church. I am thinking here of a congregation of 200 or more meeting in the church or church hall where a domestic television set would be unsuitable. A projection television can be used successfully for a meeting of this size, but it does cost quite a lot to hire. If the church is planning to make regular use of video in its services and other activities, it should consider investing in a projection set. Otherwise one could consider the linking together of a number of domestic television sets placed high enough off the ground for them to be visible over the heads of the people in front, and use this Heath Robinson method to show the video presentation. One church in my own area did this successfully with *Chariots of Fire* and have repeated the exercise.

Among the programmes that might be considered for church viewing are the set of Mission England videos which contain the whole of eight rallies from Billy Graham's 1984 crusade, or the Science Adventure or Sermons from Science series from Fact and Faith Films (although as the latter are also available on 16 mm film, you may have the necessary projection equipment and prefer to take advantage of the superior quality of film).

There can be little doubt that the best use of video can be made in the small group situation. Picture quality decreases rapidly when there is an increased size. While large screen video equipment does exist and can be obtained at a fairly reasonable cost the quality does not stand comparison with film projection. This position will change in the next few years with the advent of high resolution cameras and video projectors but at present the limitations of large screen video showings do need to be recognised. However, it is worth noting that there will be occasions where large screen video is the only practical means of presentation and of course in such situations it is better to have something than nothing.

*The video show*

Some of the most difficult, but at the same time the most rewarding and fruitful, outreach work is that directed towards teenagers. Over the last 20 years the concept of the 'evangelistic gospel concert' has been developed and widely used amongst this age group, but it is not without its problems. Firstly, with the diversity of styles of pop music and strong devoted followings for folk, soul, reggae, rock or heavy metal, it is not easy to provide a concert to please more than a proportion of the potential audience; others get easily bored.

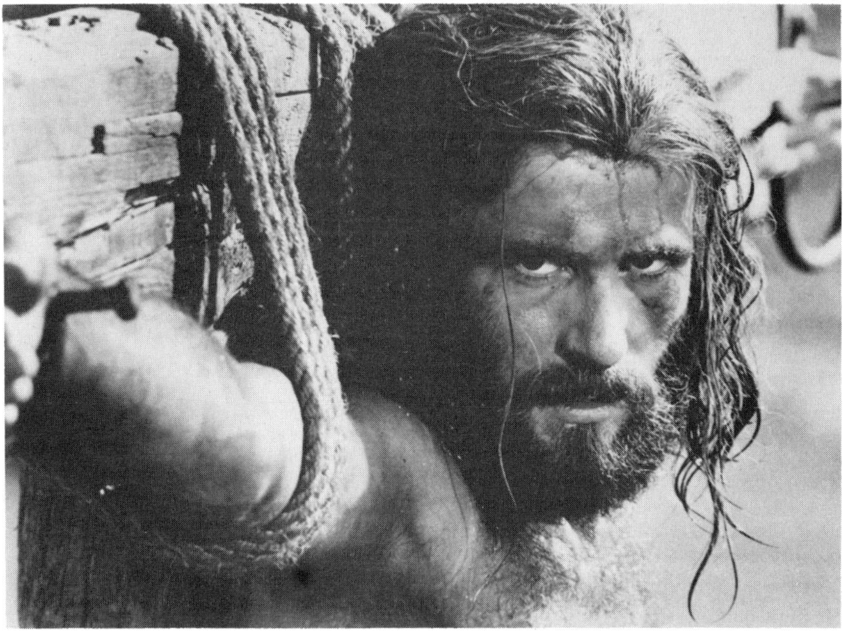

*Brian Deacon is Jesus in* The Story of Jesus *(International Films)*

Secondly, those in the performing group may be gifted musicians, but this does not always mean that they are gifted evangelists or communicators who are skilled in the presentation of a clear, consistent message. Thirdly, there remains at least a question mark over how closely the audience is actually listening to the content of the lyrics – even on the occasions when they can hear them.

The Rev. Steve Chalke believes the use of videos, both pre-recorded and live, can help us overcome some of these problems. When Tonbridge bus garage became vacant he took it over to stage *The Bus Company Video Show.* For a week, various Christian groups and artists appeared before audiences of up to a thousand. Above the stage was the 60 inch screen of a television projector, and in front were two video cameras. One camera was stationary on a tripod and the other was mobile on the shoulder of a cameraman. An operator at a monitor was able to select pictures from either camera or from pre-recorded cassettes.

The overwhelming influence of television is widely recognised. It is estimated that children over the age of 5 spend between three and five hours every day watching television, which is of course bound to condition their thinking patterns. They and we have grown used to the quickly changing images that tv presents. From *Top of the Pops* and

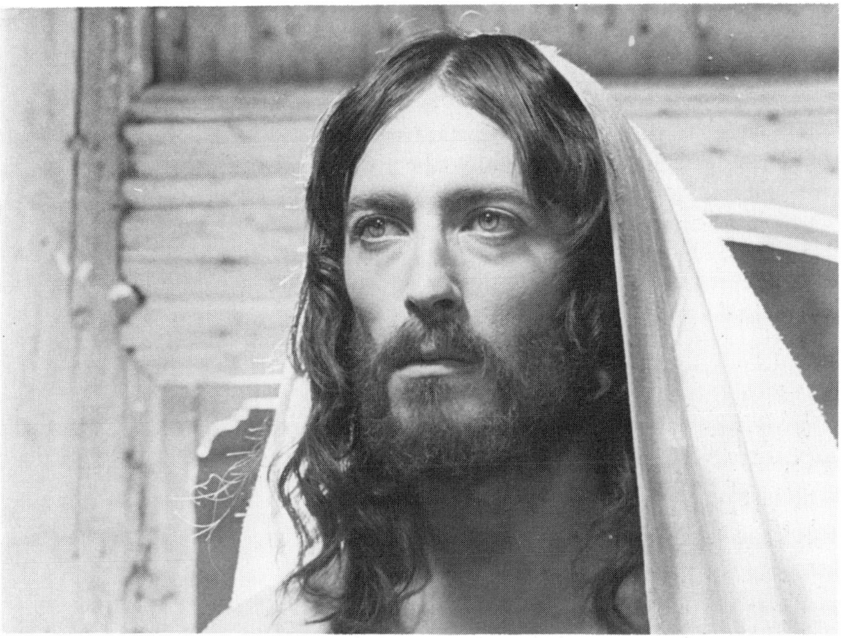

*Robert Powell plays Jesus in* Jesus of Nazareth *(Precisiosn Video)*

*The Tube*, to *News at Ten* and the chat shows, all employ the same device – fast changing pictures and different voices. At The Bus Company Video Show, through the use of the large video projector and cameras, it was possible to present the same kind of fast moving and stimulating menu that television provides.

The group on stage performed one or two songs with lyrics appropriate to the theme of the evening, with the soloist or instrumentalist visible in close up on the screen, then there was projected an interview with a pop star which had been previously recorded or a pre-recorded video from Christian artists such as Sheila Walsh or Mark Williamson. After more live music in a different style there might be an interview on stage, again viewed through either the fixed or mobile camera. Mime and drama were also used. Each evening's presentation would end with a 10–15 minute evangelistic address based on the theme of the whole programme and drawing on some of the images seen during the show.

Using video for evangelism amongst teenagers in this way has several benefits. Firstly, because the show is fast moving the audience does not get bored. Even if they don't like a particular element which is included in the programme such as a couple of songs from a rock group (whether live or on video) they understand that soon the image will be changed and the next element in the programme will be different. Secondly, there are many amateur Christian groups who, whilst being able to perform perhaps four or five songs very well during the course of an evening, are overstretched when asked to present a whole concert. This format therefore offers them the opportunity of performing without pushing them beyond their capabilities. This principle also applies to the use of drama and mime groups. Thirdly, the format gives the opportunity to develop a particular theme and ensure that the evening really does have a lot of meaningful content which is all leading in the same direction. For instance, one evening in Tonbridge the theme of 'What's Gone Wrong With Our World?' was used, and each group or musician taking part, along with those involved in drama as well as the preacher, understood that their material should be related to this theme. So through the evening a foundation was laid and slowly built upon block by block to the stage where at the end of the evening a response could be called for.

The proof of this formula lies in the results. The show ran from Monday to Saturday and by the end of the week more than 900 were being admitted and several hundred had to be turned away. Half the audience were youngsters on the periphery of church activities, in youth groups, etc. The other half had no Christian connection but had been attracted by publicity in schools during the previous week or by the free bus service that ran each evening from surrounding towns and villages. 120 conversions were recorded of which 60 per cent were known to be still within the local churches after a year.

Similar presentations have been used at Spring Harvest and at Greenbelt. There is a growing number of music videos now available from Christian producers covering a wide range of musical taste. Many are straight recordings of concerts, but some are more akin to pop videos in their compilation. Several secular groups, such as Kajagoogoo, have videos that lend themselves to use in Christian outreach situations.

### *Copyright*

It is appropriate to repeat here a reminder about copyright restrictions. While videos from religious producers will probably be made available without any limitation on the way in which they can be used, others – particularly those that are derived from feature films – are intended only to be used for domestic viewing, and any other use requires a licence to be obtained from the copyright owner or his agent (who is usually the national distributor of the film). An exception is, however, made for performances in schools.

## Video relay meetings

1984 and 1985 saw the culmination of a major evangelistic effort known as Mission England. This was the umbrella for a number of crusades led by Dr. Billy Graham and it extended the concept of video relay meetings. The Billy Graham team used video in two main ways to increase the audience both simultaneously and subsequently.

It provided an overflow facility from the football grounds in which Dr. Graham was speaking. These overflow meetings were held in the open air at the same time as the main crusade meeting, and the signal was relayed by means of an infra-red transmitter to a large screen video display. These very large displays operate on a different principle from a television screen. Instead of a fluorescent screen being scanned by an electronic beam, there are myriads of light bulbs whose brightness (or lack of) form the picture. To appreciate the picture, one needs to be at least 60 feet away from the screen, which is about the size of the side of a large furniture van. Nearer than this one is aware of the clusters of light bulbs so that the quality of the picture is reduced.

While the sense of participation in the meeting is reduced, one will probably get a better view of the speaker than will somebody looking from the terraces across the pitch to a very distant stage.

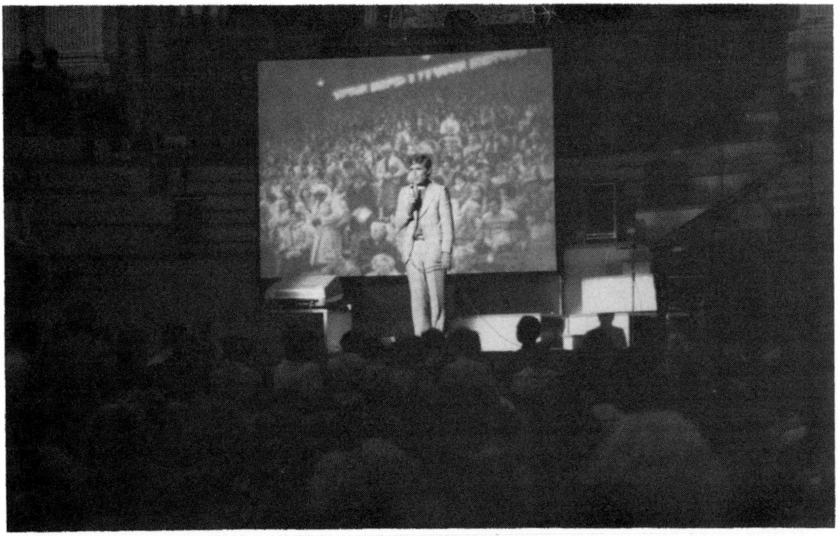

*Bishop Pat Harris stands in front of the screen as people respond to the appeal at the end of a video relay meeting of Mission England in Huddersfield Town Hall. In Yorkshire there were some 40 separate video presentations in ten public locations, which attracted 23 515 people. Of these, there were 1196 responses and 686 known decisions for Christ. 215 of these were at Huddersfield (Gospel Vision)*

The second way in which Mission England used video was in their 'delayed video missions'. A number of the stadium meetings were recorded by outside broadcast units, the resulting programmes were edited and duplicated, and they were sent away to many different parts of the country. The main advantage of using video in this way is that many people who would not have been able to get to a live Billy Graham meeting because of the travel distances involved were able to listen to his preaching. The Mission England organisation have produced comprehensive checklists as guides to setting up and running such meetings.

I have had an opportunity to talk to various people involved in this project from the organisers at the national level down to people in local churches at the receiving end of the service. There can be no doubt that by and large this attempt at making video evangelism available was well received but of course not without its critics. The major criticism that has been levelled at such meetings is the lack of atmosphere and feeling of involvement when watching something one knows was recorded many miles away at a different time.

Other complaints have related to picture quality, because in spite of all the advantages of speed and convenience of video projectors there is no doubt that the resulting enlarged picture is poorer than we are used to with 16 mm films or Cinemascope at our local ABC. The picture quality is also noticeably poorer than the domestic television set.

In spite of these limitations, there are circumstances which present opportunities for evangelism that will be within the scope of the local church or local interdenominational committee.

Some mission committees have taken over local cinemas for a week's showing of a film like *Jesus.* For the most part this has been done by arranging for a cinema quality copy of the film to be available, but if your local cinema has video projection facilities (and some do) then the scope is greater and you may be able to consider a video programme such as *Jesus Then and Now.*

My own general feeling is that for large screen evangelistic presentations I would rather use film than video. I have two main reasons: firstly, the picture quality is better, secondly, there is so much more good film material. Some distributors of good Christian films are listed at the end of this chapter.

## Conclusion

I have reviewed some of the ways in which video is being used in evangelism. You may think of many other possible applications, but in this book we are seeking to demonstrate proven ways in which the new medium can be used. The advent of the video recorder has placed in our hands a tool which, if we have our eyes open, will enable us to take the initiative in taking the gospel into people's homes. I have recently considered 'cold calling' on my neighbours with a piece of literature explaining the content of *In the Bin* by Scripture Union, saying that I am concerned to be able to talk about the Christian faith to people about me and I wonder whether they would permit me to show them a programme so that we might spend half an hour one evening discussing the Christian faith. In most cases I will find that I don't have to take my own video with me, I can just turn up at their home with the cassette, a piece of evangelistic literature and my bible. Five years ago I wouldn't have been able to undertake this exercise without carrying a projector and a screen around in my car which would have taken at least a quarter of an hour to set up and 10–15 minutes to take down. As it is I can now enter a home without overwhelming the occupants with the gear that I bring with me. Please let me plead with you to think creatively about how you are going to share the love of God with those around you.

## Training for evangelism

The need to understand how to evangelise is so significant that I want to draw attention to four programmes recorded during Spring Harvest's early ventures into programme making. They are entitled *Radical Evangelism* and were recorded during seminars given by Roger Forster and Harry Sprange in the Big Top during Spring Harvest 1981. Their technical production leaves much to be desired but their content is superb. The programmes focus on a number of aspects of evangelism including 'The Enemy and our Weapons', 'The King and His Troops', 'The Invasion and the Plan'.

These programmes help one to understand the difference between preaching to convince the mind, to bend the emotions or to affect the will. The problem of emotional blackmail and over-pressuring people into making decisions is also very helpfully discussed.

Direct training in evangelism is provided by a series of videos produced by Campus Crusade for Christ under the title *Way of Life Training* and presented by Nick Page amongst others. This very helpful series of programmes not only gives the theory of evangelism, but also through acting demonstrates the 'how to's and 'how not to's when approaching friends and workmates with the gospel. Both the Spring Harvest and the Campus Crusade material come complete with seminar notes.

## Teaching on evangelism

| | | *duration* | *distributor* |
|---|---|---|---|
| Roger Foster & Harry Sprange: Radical Evangelism | four tapes, each | 1:00 | Spring Harvest |
| Clive Calver: I Believe in Evangelism | | 0:36 | Spring Harvest |
| Peter Meadow: How to Work with Young Marrieds | | 0:36 | Spring Harvest |
| Way of Life Training: | | | |
| 1. The Cleansed Life | | | |
| 2. How to be Filled with the Holy Spirit | | 1:40 | CCC |
| 3. How to Walk in the Spirit | | | |
| 4. How to Witness in the Spirit | | 1:40 | CCC |
| 5. The Great Commission | | | |
| 6. How to Love by Faith | | 1:40 | CCC |

## Videos for evangelism

| | | | |
|---|---|---|---|
| Number One; The Stranger; In the Bin | | 0:41 | SU |
| Science Adventure Films: | | | |
| A Mystery Story | | | |
| Clown Faced Carpenter | | | |
| A Matter of Taste | each part | 0:10 | FFF |
| Journey to the Stars | | | |
| The Wonder of You | | | |
| Water, Water Everywhere | each part | 0:10 | FFF |
| Sermons from Science: | | | |
| City of the Bees | | 0:30 | FFF |
| Where the Waters Run | | 0:30 | FFF |
| The Ultimate Adventure | | 0:30 | FFF |
| The Story of Jesus | | 1:50 | Thorn-EMI |
| Eric Delve: Why I Believe the Resurrection is True | | 0:45 | Spring Harvest |
| Reinhard Bonnke: Where are Thy Accusers? | | 0:45 | AOG |
| Terry Hanford: The Touch that Means so Much | | 0:22 | AOG |
| Luis Palau: Businessmen's Breakfast | | 0:40 | Gospel Vision |
| Luis Palau: Video Evangelism | three talks, each | 0:10 | Miss Lon |
| Luis Palau: An ABC for Evangelism | three talks, each | 0:07 | Gospel Vision |

| | | | |
|---|---|---|---|
| Peter Anderson: The Relevance of Christianity Today | five talks, each | 0:15 | CTA |
| Billy Graham: Bristol programmes | four tapes, each | 1:30 | Miss Eng |
| Billy Graham: Birmingham programmes | four tapes, each | 1:30 | Miss Eng |

## Music videos

| | | | |
|---|---|---|---|
| Dave Bilborough at The Banquet | | 0:25 | MKMCV |
| Tensing-Bergen Youth Choir: Fun and Faith | | | Vision |
| Johnny Cash: Gospel Road | | | World Wide |
| Greenbelt Live | | 0:50 | Grenville |
| Maranatha Music concerts | three tapes, each | 1:00 | MKMCV |
| In Concert with Noel Paul Stookey | | | Bagster |
| In Concert with Barry Crompton | | | Bagster |
| In Concert with Steve Flashman | | | Bagster |
| In Concert with Adrian Snell | | | Bagster |
| In Concert with Mark Williamson Band | | | Bagster |
| In Concert with Servant | | | Bagster |
| Sandi Patti in Concert | | 0:54 | Bagster |
| Kids Praise 4 | | | Word |
| Amy Grant in Concert | | | Word |

## Testimonies

There are many individual testimonies on video that might, with careful selection, meet a specific evangelistic need. Most are from small Christian video production companies, and all are listed in *Guide to Christian Video*.

## Film Distributors

| | | |
|---|---|---|
| Christian World Centre | PO Box 30<br>123 Deansgate<br>Manchester M60 3BX | 061 834 6060 |
| CTVC | Beeson's Yard<br>Bury Lane<br>Rickmansworth<br>Herts WD3 1DS | 0923 777933 |
| Grenville Film Productions | Kedgeworth House<br>Blandford Heights<br>Blandford<br>Dorset DT11 7TE | 0258 56303 |
| International Films | 235 Shaftesbury Avenue<br>London WC2H 8EL | 01-836 2254 |

## Video distributors

| | | |
|---|---|---|
| Assemblies of God | 14 Worths Lane<br>Haughton Green<br>Denton<br>Manchester M34 1GB | 061 336 8971 |
| Bagster Video | Westbrooke House<br>76 High Street<br>Alton<br>Hants GU34 1EN | 0420 89141 |
| Campus Crusade for Christ | 103 Friar Street<br>Reading<br>Berks RG1 1EP | 0734 589461 |
| Christian Television Association | 27 Ash Hayes Drive<br>Nailsea<br>Bristol<br>Avon BS19 2LQ | 0272 856447 |
| Fact and Faith Films | 120 The Rock<br>Bury<br>Lancs BL9 0PJ | 061 764 1538 |

| | | |
|---|---|---|
| Gospel Vision | 143 Toller Lane<br>Bradford<br>West Yorks BD8 9HL | 0274 41437 |
| Mission England | 27 Camden Road<br>London NW1 9NL | 01-267 2346 |
| Mission to London | 112 City Road<br>London EC1V 2NB | 01-251 6534 |
| MKM Creative Vision | M&K Microtech Ltd<br>131 Gratton Road<br>Bradford BD1 2HS | 0274 306089 |
| Scripture Union | 130 City Road<br>London EC1V 2NJ | 01-250 1966 |
| Spring Harvest | Buena Vista<br>South Hanningfield Road<br>Wickford<br>Essex | 0268 710026 |
| Thorn-EMI Video Programmes | Thorn-EMI House<br>Upper St Martin's Lane<br>London WC2H 9ED | 01-836 2444 |
| Vision Productions | 111 Ecclesall Road South<br>Sheffield S11 9PH | 0742 367 3318 |
| Word (UK) | Northbridge Road<br>Berkhamsted<br>Herts HP4 1EH | 04427 74711 |

7

# A video ministry

## An account of work in the scattered communities of the Scottish Highlands

We have considered in the previous chapters various ways in which a church may, within itself, use video to educate and train its members, nurture its children, put the claims of the gospel of Christ before its friends, and go out to present an evangelistic challenge to the secular world. As should now be clear, all these activities depend for their success on the presence of an informed leader, teacher or presenter. Many of the videos that have been mentioned have been designed specifically to be used in a structured way under the guidance of a group leader. These leaders will be found within the fellowship of larger congregations, while a smaller community may depend upon its minister.

But now let us look at the way in which video can 'fill a gap'. There are many very small groups of worshippers, often elderly and in isolated areas, who do not have a full-time leader, who perhaps do not have a leader at all. There is now no need for such a group to go short of sound Christian teaching. The world's finest preachers and teachers can be packaged in a 7 × 4 inch cassette and made available at the press of a switch.

One area with a large number of widely scattered small communities is the Highlands of Scotland. In 1980 Jim Eglinton, who runs a garage in Inverness, felt a burden for the needs of these people. His hobby was photography and cine filming, and he was impressed by the Lord that video was a tool for the future, a medium that he was going to use in a variety of situations for the extension of his kingdom. Over a period, this vision from God became very clear as the tremendous potential of the medium was realised.

Initially it was envisaged as a means whereby Christians who lived away from the urban centres and off the beaten tracks would have the opportunity to hear (and see) preachers who, due to the distances involved, would normally be unavailable, and in parallel with this would be a tool for evangelism. While these aims have subsequently been fulfilled, the vision has also been enlarged. As people have come to realise what resources are available in the video equipment, so many varied requests have been received to meet a diversity of needs.

When the Highland Christian Video Fellowship was formed it had no available finances. Jim knew of no one involved in Christian video or who could share his vision. Then to his surprise and delight he stumbled upon a stall at the Keswick Convention advertising the work of a Christian video ministry based in Edinburgh. This provided much needed encouragement.

Some two weeks later, a gift was received of a video camera and recorder. Without any delay, the initial recording attempt was made at a local evangelistic rally. The excitement on the faces of the participants as they viewed themselves on the screen was indescribable!

The pace at which HCVF grew after those early days was quite remarkable. The necessary funding became available, some from very unexpected sources. Equipment of a high standard was given. Jim travelled extensively to record events, gratified to see the tapes being distributed and well received. Premises in two local church properties were made available, but they were speedily outgrown by the recording needs.

In those early days, the tapes were distributed on a free loan/donation basis, but it soon became evident that things had to be organised more efficiently. The HCVF membership scheme was introduced, each member paying a one-off subscription of £10 and the subsequent hire charges. It was felt that the idea of a video club would foster a more responsible attitude in those who were using the tapes, would encourage subscribers to use the tapes more extensively, and the returns would fund the purchase of more tapes. The membership presently located throughout Britain (with one or two abroad) stands well over a hundred.

While the production side of HCVF has prospered, with a tailor-made recording studio and high-quality equipment, we are more concerned in this account to show how pre-recorded material can be used to God's greater glory. Not that the two activities can be divorced from one another. Many of the pre-recorded programmes that are used by HCVF are ones that they themselves have made.

Having local settings or other associations, these can be particularly effective.

Among programmes offered by HCVF which have been used beneficially by them are four home bible study tapes on *Jonah* by the Rev. Murdo A MacLeod of Greyfriars Free Church, Inverness and a testimony by Rev. Fergus Robertson, minister of Dalneigh Church of Scotland, describing his conversion after seven years in the ministry.

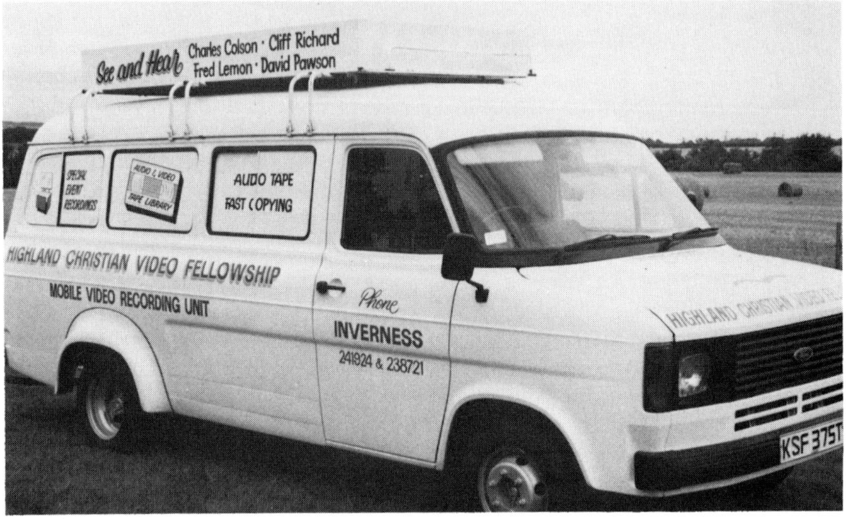


*The mobile video recording unit of Highland Christian Video Fellowship*

Most of the visitation work is done by just one person who travels out with the equipment, sets it up and operates it, then packs it up and brings it home at the end of the session. A tremendous asset has been the acquisition of a Ford Transit minibus which, painted up in the colours of HCFV, has become a familiar sight on the roads out of Inverness. The van has undergone a major overhaul and also doubles as a mobile video recording unit.

It might be thought that some of the basic equipment, like the video recorder, might already be available at the location but experience has shown that those who possess recorders are unwilling to let them be used, while those who are keen to use them do not possess them! Consequently it is necessary to take out a projection set or several standard receivers. In the early days it was common to set up four or six 26 inch television sets down the sides of a hall. Now an ITT 6 foot screen is often used.

Feature films always prove popular. Two well used ones are *The Cross and the Switchblade,* telling the dramatic story of David

Wilkerson's ministry among the gangs of New York and the resulting conversion of gang leader Nicky Cruz, and *The Hiding Place*, based on Corrie Ten Boom's account of the tribulations of her family while helping Jews in Nazi-occupied Amsterdam. Another programme that has been used to bring blessing is *From Witchcraft to Christ*. This is the testimony of Doreen Irvine who wrote the book of the same name. It was filmed by HCVF when the author visited Inverness, and has subsequently been widely used locally and elsewhere.

Video programmes have been used to advantage in churches in place of a 'live' sermon. There are many sermons listed in the *Guide to Christian Video*, but many are too long for use in a normal Sunday service. There has been a whole spate of one and even two hour (or longer!) 'head and shoulders' programmes. While these have a valuable teaching role, they are of limited value for day to day use.

Half an hour is probably the optimum length for a programme to be used in the context of a church service. Two excellent series have been issued by Bagster Video that each consist of five tapes each 30 minutes in length. In *I Have Decided to Live Like a Believer*, Tony Campolo deals with relationships within the family and outside, with individual titles such as 'Living with Yourself', 'Living with Your Spouse', 'Living with Your Family', 'Living in the Community', and 'Living in the World'. The second series is a set of five studies in II Corinthians by Luis Palau, entitled *Victorious Living and World Evangelism*.

A number of videos by Eric Clarke on various cults (including *Jehovah's Witnesses, Mormons* and *Moonies*) have also been used successfully, although these are rather longer. These were shot by HCVF and are available from them or from Eric Clarke at Christian Information Outreach. All these programmes stimulate subsequent discussion.

Church services are not the only situation where video tapes can be usefully used. One of the local churches pursued the idea of introducing video suppers whereby they divided their parish into areas, each inviting their neighbours and contacts to watch Christian videos, thus using video as a tool for evangelism. (This has been described in detail in the previous chapter.) The effectiveness of this method cannot be over-emphasized. Christians should be constantly encouraged to participate in coffee mornings, video evenings, etc., as a means of outreach.

People are geared to visual presentation in this era of television. Education is reinforced by visual aids. The media in general capitalise on visual effects and it is helpful for the church to understand this trend when we explore new avenues of communication. Many people will watch a video when they will disregard an audio presentation.

In this context of evangelistic outreach, HCVF have supplied numerous tapes to schools, SU groups, youth camps, rallies, youth fellowships, old people's homes, prisons, Sunday schools and to individuals throughout the country showing the tapes at various venues and functions, as the opportunities present themselves. The potential is unlimited. For example, it is easy for an average SU group to organise a school outreach using the medium of video as most schools have access to a video recorder. Suitable speakers are not always available whereas the cost of hiring a video would normally be well within their budget.

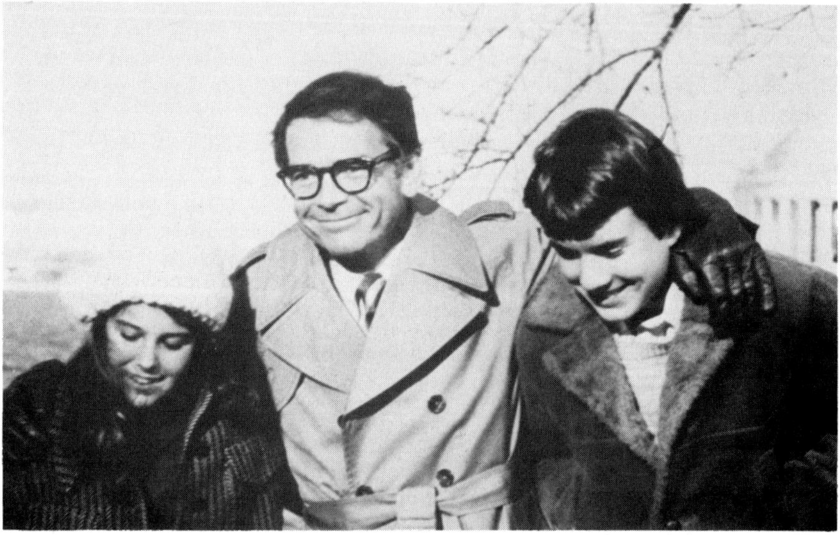

*Charles Colson greets his children after his release from prison, as seen in* Born Again *(Embassy)*

The local prison holds a Sunday afternoon service for short-stay inmates. Video programmes used on these occasions include *Born Again,* which is the dramatised story of Charles Colson who was himself imprisoned in the aftermath of the American Watergate scandal, was converted, and who has subsequently devoted himself to the service of men and women confined in prisons. Two further programmes featuring Charles Colson are available from HCVF. They were filmed at meetings which were addressed by Colson when he visited Scotland in 1982. Another local production which has proved of value for such meetings is the testimony of Fred Lemon, himself an ex-convict and author of two best-selling books – Breakout and Going Straight. This programme, entitled *From Crime to Christ,* was filmed when Fred was speaking at a meeting at Inverness Town Hall.

It is difficult to get much feedback or follow-on from this prison ministry, because the prisoners involved are all short term. But there have been many subsequent requests for bibles, and who can estimate the extent of the influence that these will have, by God's mercy, amongst many who might never have heard the challenge of the gospel but for their stay in prison.

Many believers have testified that they have been wonderfully blessed and encouraged, challenged and inspired, through the availability of some of these tapes. There is something wonderful about sitting down of an evening to watch and hear Corrie Ten Boom speak about *The Lives She's Touched,* to mention but one of many

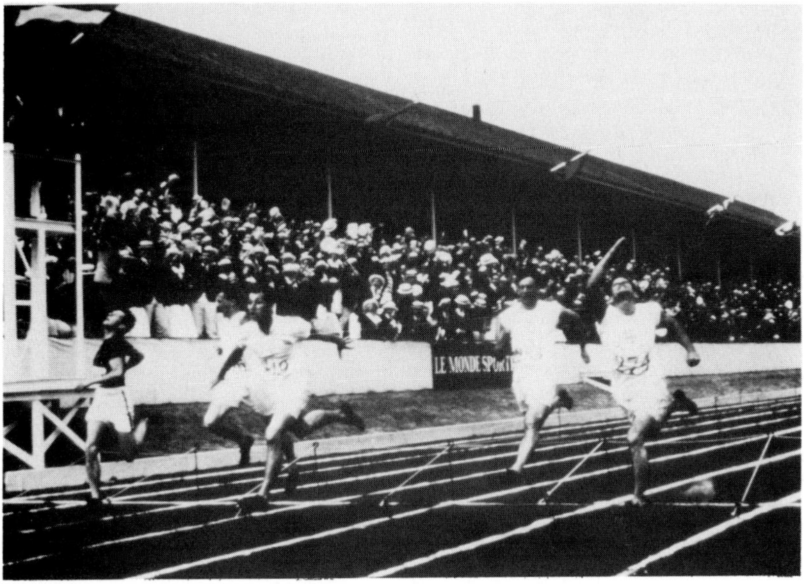

Chariots of Fire *has been widely used in outreach ministry (CBS/Fox Video)*

challenging films now available. Many have proved the bible study teaching tapes to be a great help.

In a vision for the future is the acquisition of a coach. A larger vehicle that could be converted to have a video lounge with monitors and enough chairs to seat a school class, etc., would have enormous potential in an area such as the Highlands. No longer would the video message be confined to groups with suitable buildings. A video ministry based on such a vehicle could be used in small villages and schools, taking Christian programmes and showing them to many who would otherwise never hear good sound bible teachers or potent personal testimonies. And of course the ministry does not have to be restricted to the serious and the earnest. There are a growing number of music videos and, particularly for children, cartoon and adventure programmes that carry their message in a joyous and entertaining way.

Such a ministry is applicable not only in the Highlands. There are many rural areas in England, Scotland, Wales and Ireland that would be fertile ground for such a mission to small villages. This should be perhaps a challenge to some of our more affluent town congregations.

Jim Eglinton comments: "God has led us to this point, and we are encouraged to believe that he will continue to use this ministry which he has inauguraged and equipped. We praise him for what he has done and we give him all the glory.

"We ourselves would like to see more tapes being distributed to the house-bound, both Christian and non-Christian. Isolated believers, who lack the support of a local fellowship would benefit greatly from the use of Christian video tapes. We would also like to see a breakthrough into hospitals. These are goals for the future. The possibility of an itinerant ministry through mobile video showbuses, to the rural and isolated communities, may sometime become a reality.

"Meanwhile we praise God for placing this tool in our hands, as we anticipate great things for his glory by sowing the seeds of the kingdom and by building up the church."

## Videos mentioned in chapter

| | | *duration* | *distributor* |
|---|---|---|---|
| Jonah | four tapes, each | 0:30 | HCVF |
| Conversion of a Minister | | 0:30 | HCVF |
| The Cross and the Switchblade | | 1:45 | Gateway |
| The Hiding Place | | 2:05 | Bagster |
| From Witchcraft to Christ | | 1:20 | HCVF |
| I have Decided to Live Like a Believer | five tapes, each | 0:30 | Bagster |
| Victorious Living and World Evangelism | five tapes, each | 0:30 | Bagster |
| Jehovah's Witnesses (two parts) | | 2:05 | CIO/HCVF |
| Moonies | | 1:00 | CIO/HCVF |
| Mormons | | 1:00 | CIO/HCVF |
| Born Again | | 1:50 | Embassy/ Bagster |
| Charles Colson at Dungavel Prison | | 0:40 | HCVF |
| Prison Christian Fellowship (Charles Colson) | | 1:00 | HCVF |
| From Crime to Christ | | 1:10 | Videout |
| The Lives She's Touched | | 1:00 | Bagster |

## Some distributors

| | | |
|---|---|---|
| Audio Visual Ministries | 152 High Street<br>Holywood<br>Co Down BT18 9HT | Holywood 4577 |
| Bagster Video | Westbrooke House<br>76 High Street<br>Alton<br>Hants GU34 1EN | 0420 89141 |
| Christian Information Outreach | 92 The Street<br>Boughton<br>Faversham<br>Kent ME13 9AP | 0227 751489 |

Gateway Films are available from Bagster Video

| | | |
|---|---|---|
| Harvestime Video Library | Harvestime House<br>136 Hall Lane<br>Bradford<br>West Yorks BD4 7DG | 0274 729752 |
| Highland Christian Video Fellowship | 46a Seafield Road<br>Inverness | 0463 224788 |
| Kingsway Video | Lottbridge Drove<br>Eastbourne<br>East Sussex BN23 6NT | 0323 27454 |
| Videout | Carrubbers' Christian Centre<br>65 High Street<br>Edinburgh | 031 667 3318 |

# 8

# Religious entertainment!

Ken Jackson

Contrary to some opinions, and in spite of the apparent evidence of some forceful advertising, there is a wide range of video programmes available through the general trade that is worth viewing and which may form acceptable entertainment for a Christian family.

It would not be appropriate to try to itemise here all the videos that lend themselves to family viewing. A list would soon become out of date as new titles are released and old ones are withdrawn from sale. A catalogue of such programmes has been compiled by Trinity Video and can be obtained from them.

Instead, I would like to pick out videos from the general secular distributors that have a biblical theme or relate to a topic of Christian interest.

The ones that have probably been around the longest are the *Greatest Heroes of the Bible* series from VCL. Although these were dropped from the VCL catalogue in 1984, many Christian video libraries still hold copies. There are 14 tapes in all, each produced originally for American television. If you do not mind your bible heroes presented with a touch of razzamatazz and talking with American accents (as they do in the most of the films that are going to be mentioned here – which is after all as valid as them having English accents) then these films provide dramatic versions (with a little extra added excitement) that will make the old testament stories live for junior church classes. They range from the story of Noah in *The Deluge* through Abraham, Lot, Joseph, Joshua, Samson, David,

Solomon and Daniel to Esther. Queen Esther is played by Victoria Principal (who subsequently graduated to 'Dallas').

Three of these 'heroes' are also the subjects of full-length films: *Moses* (Precision Video) paints a vivid picture of the trials of the Children of Israel in Egypt. Burt Lancaster and Anthony Quayle are a rather benign Moses and Aaron. The photography is excellent and if the characters of the principals do not come over as we expect, perhaps our expectations are wrong! This film, like many of the others, is a useful challenge to our preconceptions. It is also a timely reminder of the authority and majesty of God. Cecil B. DeMille deals with the same subject in *The Ten Commandments* (CIC Video). This film has been hailed an an all-time classic, particularly for its theatrical effects. The parting of the Red Sea and the destruction of the golden calf are masterpieces of film-making ingenuity. Charlton Heston's performance as Moses matches the occasion.

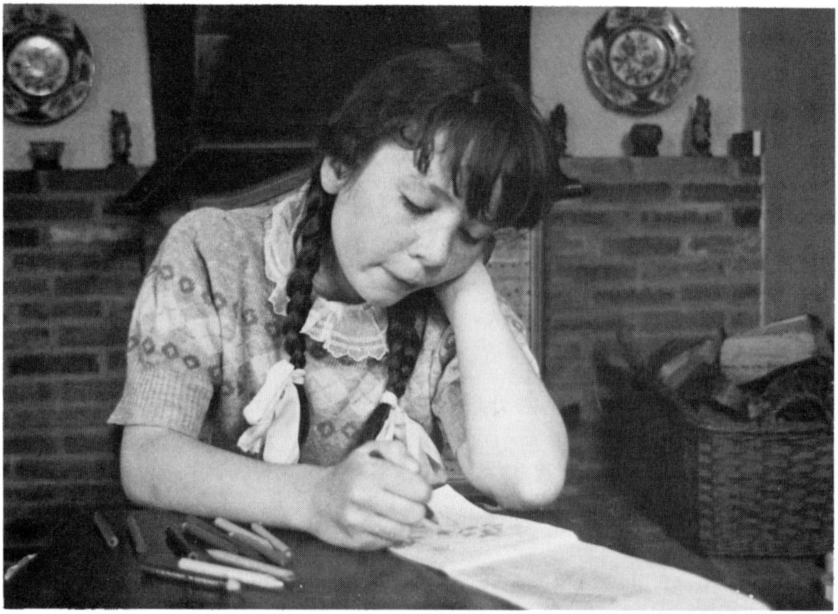

*Ruth in a thoughtful mood, in* Tanglewoods' Secret *(International Films)*

David gets more cavalier treatment in *David and Goliath* (MEVC). In fairness, it is stated that "the screen play is freely adapted from the bible". Some parts of the storyline will seem familiar, but a dialogue bristling with 'thees' and 'thous' appears to be an attempt to make it seem more authoritative than is justified. The film was shot on location in Jerusalem and Yugoslavia.

Precision Video have released *Samson*, but this is *not* the biblical character. *Samson and Delilah* from CBS/Fox however is (more or less) true to the account in Judges.

The biblical story of Lot is barely discernible in *Sodom and Gomorrah* (VPD). Stewart Grainger, Stanley Baker and Pier Angeli do their best and the film contains enough violence to warrant a PG certificate but this is not one for the Sunday school!

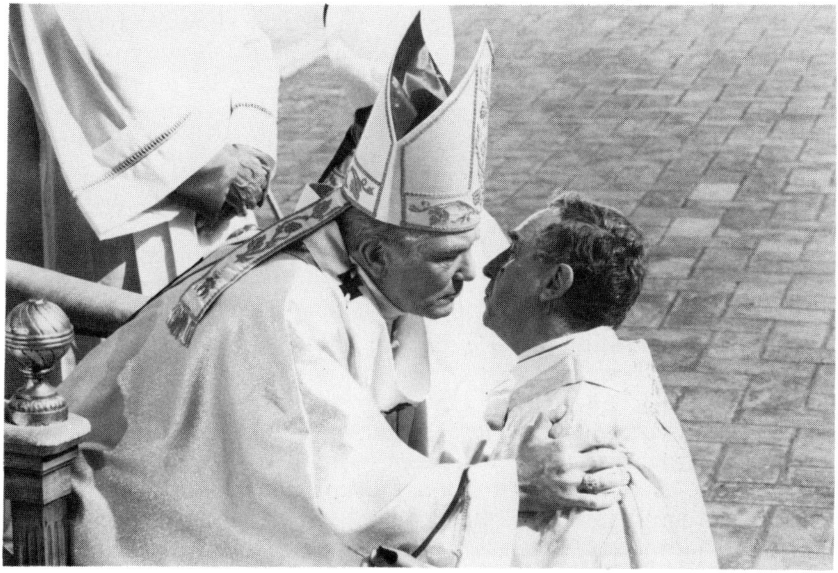

*The Pope (Albert Finney) embraces Stefan Cardinal Wyszynski (Nigel Hawthorne) of Warsaw in St Peter's Square on the day after his election. A scene from* Pope John Paul II *(CBS/Fox)*

*The Bible* has been reissued by CBS/Fox and you may get *The Old Testament* from Precision. The latter title is misleading as the film tells the story only of the Maccabean rebellion in the second century BC.

*The Robe* (CBS/Fox) is based on the novel by Lloyd C. Douglas and tells of the effect that the robe of Jesus has on the Roman soldier who wins it with his dice at the foot of the cross and on the lives of others connected with him. Richard Burton is the soldier, Jean Simmons his girl friend and Victor Mature his slave. Although the techniques of this film now seem rather dated (it was made in 1953) it has a real value for evangelism.

Two classic films must be mentioned from MGM/UA. *Ben Hur* must rival Ten Commandments for the title of most spectacular motion picture; it is perhaps best known for its breath-taking chariot race. Charlton Heston again plays the title role. *Quo Vadis*, made in 1951, tells of the establishment of Christianity in Rome. We see the burning of Rome by Nero (a convincingly mad Peter Ustinov) while Robert Taylor and Deborah Kerr provide romantic interest as early Christians. Both Peter and Paul make brief appearances and a long sequence details the martyrdom of the Christians in the Colosseum.

Two lesser films (from MEVC) are *Constantine and the Cross*, an action-packed story of the establishment of Christianity as the state religion of Rome in 300AD, and *Joan of Arc*, the tragic story of the Maid of Orleans whose devotion to God gets entangled with the political and military ambitions of others. Ingrid Bergman won an Oscar for her performance.

While film-makers of the 1950s and 60s concentrated on old testament events and Roman epics, recent films have focussed in detail on Jesus and on more recent church history. This is not the place to discuss the massive New Media Bible project which plans eventually to make a film version of the whole bible available in a variety of translations and languages. However a spin-off from it is available as *The Story of Jesus* from Thorn-EMI. The dialogue follows closely the gospel of Luke and the scenery is splendid. Inevitably the director's visualisation of the events in Jesus' ministry will differ from some viewers' preconceptions, but the whole film can be used for teaching and challenging. Equally challenging and even more dramatic is the four-part *Jesus of Nazareth* (Precision Video) made by Franco Zeffireli on location in North Africa with Robert Powell as Jesus. This is a most professional production which avoids trick effects and emphasises people's motives.

Very different in concept, but as impressive in its own way, is *Jesus Christ Superstar* (CIC Video). This modern rock opera sees the last weeks of Jesus' life through the eyes of Judas. It was shot in Israel and provides a sequence of visual and musical images.

Two films investigate historical truth and one looks at the outworking of prophecy. One of the former is *The Silent Witness* (Screenpro Films) which is the report of a scientific investigation into the origins and history of the Holy Shroud of Turin – the cloth that bears the outline of a body reputed to be that of Christ. The detective work is persuasively accurate and detailed, which is more than can be said for *In Search of Historic Jesus* (Video Brokers) which was filmed in Utah. *The Late Great Planet Earth* (VTC) is a dramatisation of the book by Hal Lindsay. It resembles an overlong tv documentary with gloom and doom presented alternately by Lindsay and Orson Welles.

On a lighter note are two videos for armchair travellers. *Jesus and the Fishermen* (Sussex Video) is a half-hour programme on the life of a Sri Lankan fishing village and particularly the passion play they produce each Easter. It has no particular message and is part of a series of 'multi-cultural music studies'. Readers will probably get more pleasure from *Pilgrimage to the Holy Land* (Guild Home Video). This is a collection of four Harlech Television programmes narrated by Rev Gwynfryn Lloyd Thomas. Unfortunately it does not decide if it is to be a travelogue or an historical survey and so fails to be either satisfactorily. It really comes to life only at the end when the Rev Thomas visits a kibbutz.

There are at least five programmes about the Pope. The personality of the present pontiff and the dramatic possibilities of his Polish background have stimulated film-makers. The first to be released was *Pope John Paul II in Ireland* (Rank Video). After his visit to Britain in

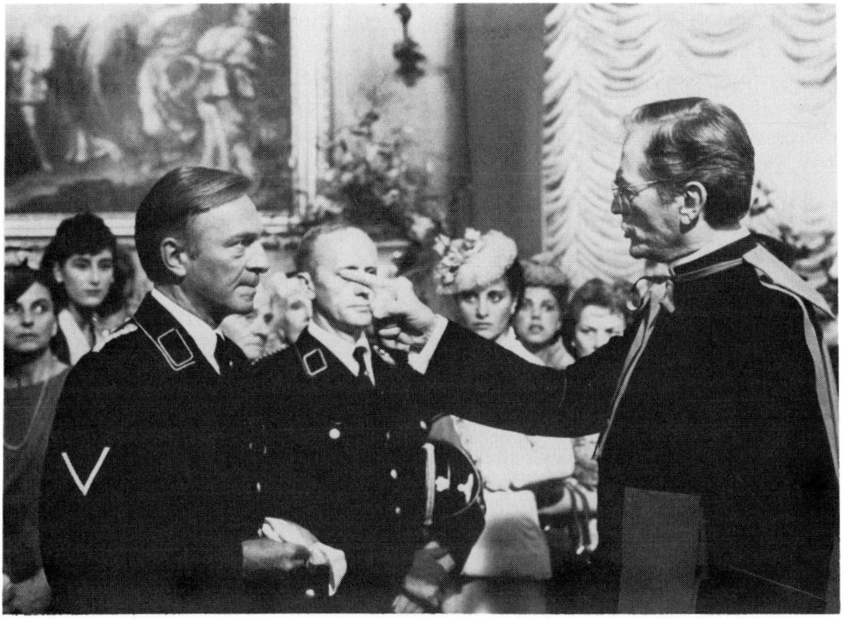

*Christopher Plummer as the Nazi commander and Gregory Peck as the priest confront one another in* The Scarlet and the Black *(Precision)*

1982, the newsreel coverage by both BBC and ITN was married together to produce an evocative record. This was issued by 3M Video as *John Paul II: The Pilgrim Pope.* At least one nonconformist was profoundly moved by this two-hour film. One man's record of part of this tour is in *The Pope in the Park* (Masterclass) which concerns the visit to Heaton Park in Manchester.

The story of the Pope's life is told in *From a Far Country* (Precision Video). It follows, with the help of newsreel footage, the events in Poland from 1926 through the German occupation and the subsequent establishment of the communist state. Karol Wojtyla was a factory worker under the Germans and a secret student for the priesthood under the communists. The same story is told in a more structured way, and with greater dramatic effect, in *Pope John Paul II* (CBS/Fox). Albert Finney plays the title role and it is a delight to observe the accuracy with which he achieves the mannerisms and speaking style of the Pope.

## Feature films

A number of feature films in the last ten years have told gripping stories about heroes of the faith. Two of these have been mentioned several times earlier in this book as useful videos for outreach and encouragement. *The Hiding Place* is set in Nazi-occupied Holland and tells the true account of the persecution of the Ten Boom family who helped Jews to escape by building a hiding place in their Haarlam house. Only Corrie survived their subsequent arrest and committal to Ravensbruck concentration camp, and the film shows her remarkable witness in prison and subsequently. *Chariots of Fire* (CBS/Fox) is the award-winning story of Eric Liddell and Harold Abrahams, who both won gold medals at the 1924 Olympic Games, and of the beliefs that motivated them.

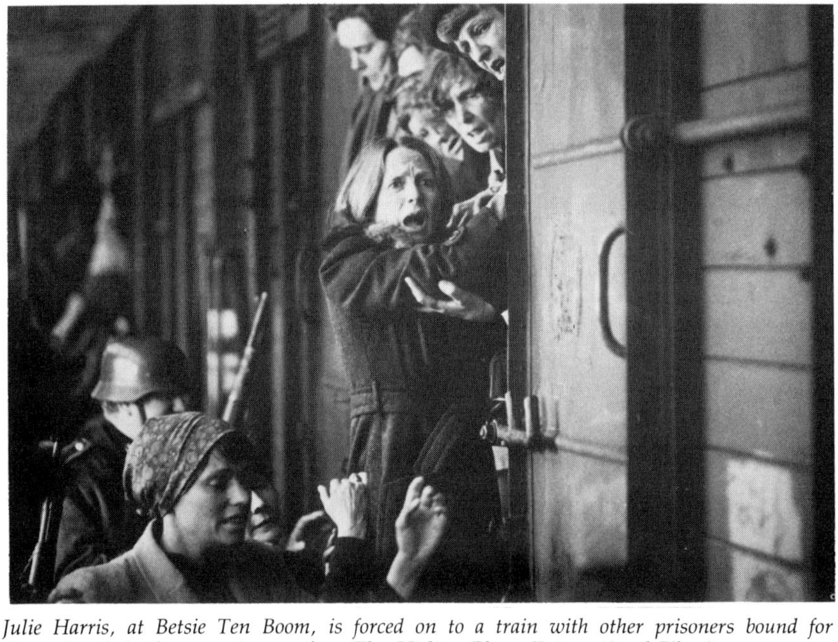

*Julie Harris, at Betsie Ten Boom, is forced on to a train with other prisoners bound for Ravensbruck death camp. A scene from* The Hiding Place *(International Films)*

Less well-known films that take religion-inspired conflict as their theme and make exciting productions include *Becket* (Odyssey) and *The Agony and the Ecstasy* (CBS/Fox). The former stars Richard Burton as Thomas à Becket and Peter O'Toole as Henry II in the clash of principles between church and state. The second has Charlton Heston as Michelangelo and Rex Harrison as Pope Julius in another clash between a dedicated artist and a warrior priest which resulted in the beauty of the Sistine chapel. The conflict in *A Man for All Seasons* (RCA/Columbia) is between Henry VIII (Robert Shaw) and Sir Thomas More (Paul Scofield). This film won six Academy awards.

A modern conflict between church authorities and secular powers is portrayed in *The Scarlet and the Black* (Precision) where Gregory Peck as a Vatican priest provides succour for escaping allied prisoners of war in Rome after the Italian surrender in 1943 and the subsequent occupation by the Nazis. *Catholics* (CBS/Fox) is subtitled 'a fable' and is the story of how one Irish priest (played by Trevor Howard) resists the abolition of the Latin mass.

In 1974, Charles Colson was sentended for his alleged part in the American Watergate scandal. *Born Again* (Embassy) contains explicit and detailed accounts of his conversion and subsequent spiritual growth. The political implications are a little difficult for a non-American to comprehend, and Colson's guilt is not clearly established. However his subsequent work for prisoners' spiritual and physical welfare is not in doubt.

## Christmas videos

As might be expected, the secular distributors have identified Christmas as a prime occasion for video buying and watching. Seasonal music is available from *Christmas Music from York Minster* (Guild) sharing a cassette with *Handel's Messiah.* The latter contains part of the oratorio sung by the Huddersfield Choral Society while the other part of the tape has a selection of musical terms by various performers. More formal music can be obtained from *Christmas Carols from Cambridge* (Thorn-EMI) which is a recording of the popular annual festival of nine lessons and carols from King's College (actually there are six lessons, read by Sir John Gielgud, and 17 carols). Music for any time of year is compiled from that popular BBC programme *Songs of Praise.* A video cassette by the same title gives a selection of hymns and congregations.

*A Dream for Christmas* (Polygram) is not really about Christmas, but it is a touching story about a black pastor and his family who have to leave Sweet Clover, Arkansas, for a church in Los Angeles. If you enjoyed the Waltons on tv you will like this story of how, against the odds, the Rev Will Douglas built up a diminished congregation and saved the building from demolition. For a Christmas present he received a bell for the empty belfry.

You might be excused for thinking that *Mary and Joseph* (Polygram) could be a Christmas story. But it isn't. It is not even a biblical story – not at least from any translation that I know!

## Children's programmes

There are several children's videos for Christmas. *Children's Tales* (Minstrel Entertainments) are two cartoon programmes from the US Lutheran Church on one cassette. Both feature Bengi and his dog Waldo. In 'Christmas Is' he is transported to Bethlehem to understand the significance of the events surrounding the nativity. In 'The City that Forgot about Christmas' his grandfather tells a tale of a city that had no love until a baby came.

Richard Chamberlain is the messenger in *The Christmas Messenger* (Videoform). Carol-singing children find the songs they are singing turned by him into animated cartoons. This is a pity really, because the choice of songs is rather eccentric and the cartoons do not improve them. A much better cartoon, and one which should please adults as well as children, is that which accompanies *Silent Night*. This well-loved carol was written by a pastor in Germany after his pipe organ had failed. The characters in the story of the attempt to repair the organ are well drawn and have their individual personalities. They offer some good moral sense.

Moral sense is the justification for including *The Selfish Giant* (Videoform) in this survey. It is based on the short story by Oscar Wilde and although not specifically Christian is often used by youth groups for their video evenings.

Also very moral, but more explicitly Christian in their teaching, are the two films from books by Patricia St. John. *Treasures of the Snow* tells how Lucien is ostracised by his friends because of an accident he caused but is befriended by an old man and gradually wins back his friendships through his selfless actions. *Tanglewood's Secret* is also a story of youthful relationships.

*God's Story* is a series of still pictures with a single voice narration constructed into a number of ten minute episodes. It was originally broadcast on Yorkshire Television, from whom the 'Old Testament' cassette can be obtained. The 'New Testament' is distributed by Guild Home Video. The whole is sympathetically and tastefully handled and presents a good encapsulation for children, in short digestible chunks, of the bible stories.

A very different presentation of bible stories is that provided by

*Superbook* from a series on American television. The word 'bible' is never mentioned, but through the medium of their 'superbook' Chris and Joy are enabled to visit the scene of, and actually participate in, such events as the garden of Eden and the flood. A series of cassettes is likely to become available.

New video programmes are being regularly released. The best of the secular offerings are likely to taken up by Christian distributors and libraries. However, if you keep an eye on the general market you may find films that are of particular interest to you or are particularly appropriate for your use. Some feature films have been mentioned that may come into this category. It is also true of pop video. A number of groups are producing programmes that include material that could be used in a Christian concert.

The range and application of video is endless. See for yourself.

## Secular distributors

| | | |
|---|---|---|
| BBC Enterprises | Woodlands<br>80 Wood Lane<br>London W12 0TT | 01-743 5588 |
| CBS/Fox Video | Perivale Industrial Estate<br>Greenford<br>Middx UB6 7RU | 01-997 2552 |
| CUC Video | UIP House<br>Beardon Road<br>Hammersmith<br>London W6 | 01-741 9333 |
| Embassy Home Entertainment | Sloane Square House<br>Holbein Place<br>Sloane Square<br>London SW1W 8NT | 01-730 3455 |
| Guild Home Video | Guild House<br>Oundle Road<br>Peterborough | 0733 63122 |
| Masterclass | Holiday Brothers<br>172 Finney Lane<br>Heald Green<br>Cheshire SK8 3PU | 061 437 0538 |
| MEVC | 122/124 Regent Street<br>London W1R 5FE | 01-434 3751 |
| MGM/UA Home Video | Hammer House<br>113–117 Wardour Street<br>London W1 | 01-439 9932 |
| Minstrel Entertainments | 61/63 Brownfields<br>Welwyn Garden City<br>Herts AL7 1AN | 07073 33716 |
| Odyssey Video | 11 Grape Street<br>London WC2H 8DR | 01-379 3441 |
| Polygram Video | 1 Rockley Road<br>London W14 0DL | 01-743 3474 |

| | | |
|---|---|---|
| Precision Video | ACC House<br>17 Great Cumberland Place<br>London W1A 1AG | 01-262 8040 |
| Rank Video | PO Box 70<br>Great West Road<br>Brentford<br>Middx TW8 9HR | 01-568 9222 |
| RCA/Columbia | 1 Bedford Avenue<br>London WC1B 3DT | 01-636 8373 |
| Screenpro Films | 5 Meard Street<br>London W1V 3HQ | 01-439 7691 |
| Sussex Video | Townsend<br>Poulshot, Devizes<br>Wilts SN10 1SD | 038 082 337 |
| Thorn-EMI Video Programmes | Thorn-EMI House<br>Upper St Martin's Lane<br>London WC2 9ED | 01-836 2444 |
| VCL Video Services | VCL House<br>9a Dallington Street<br>London EC1 0BQ | 01-251 6131 |
| VPD | Building No 1, GEC Estate<br>East Lane, Wembley<br>Middx HA9 7FF | 01-904 0921 |
| Video Brokers | 105 Heath Street<br>Hampstead<br>London NW3 6SS | 01-794 6533 |
| Videoform Pictures | Unit 4<br>Brunswick Industrial Park<br>New Southgate<br>London N11 | 01-368 1226 |
| Yorkshire Television | Television Centre<br>Leeds LS3 1JS | 0532 438283 |

# Index

action programmes, 39, 44
bible study, 24, 30
biblical epics, 103
broadcast systems, 13
cassettes, 15
children's entertainment, 110
Christmas videos, 109
church services, 83, 97
complete programmes, 35, 44
copying, 14
copyright, 20, 86, 114
discussion groups, 30, 37, 44
education, 47–61
entertainment, 70, 103–111
feature films, 108
formats, 14–15
group events, 79–82, 94
home video, 59, 73
missionary meetings, 25, 30
one-to-one evangelism, 78
pop music, 83, 91
prison ministry, 98
projection television, 12, 18–20
programme content, 20, 34
rallies, 83
recorders, 10, 13, 16–18
relays, 87
screens, 18, 19
special programmes, 40, 44
teaching, 23, 30, 64, 90
teenagers, 67, 75
testimonies, 91
training, 47–61, 88
video show, 83
video suppers, 80
women's meetings, 27, 30

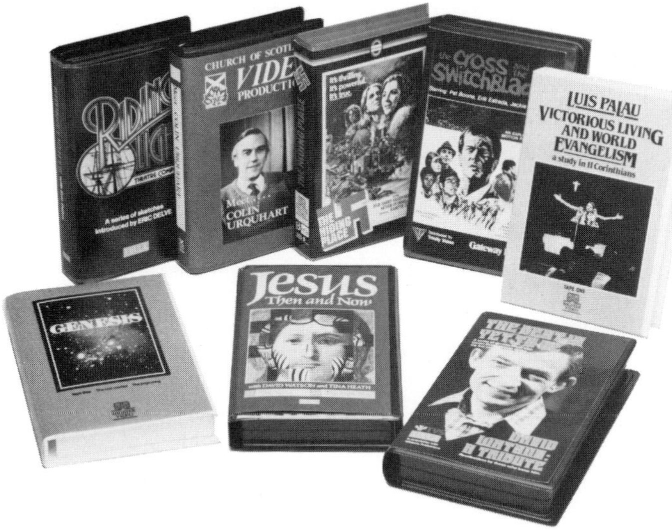
LUIS PALAU
VICTORIOUS LIVING AND WORLD EVANGELISM
a study in II Corinthians
Jesus
Then and Now
GENESIS
COLIN URQUHART

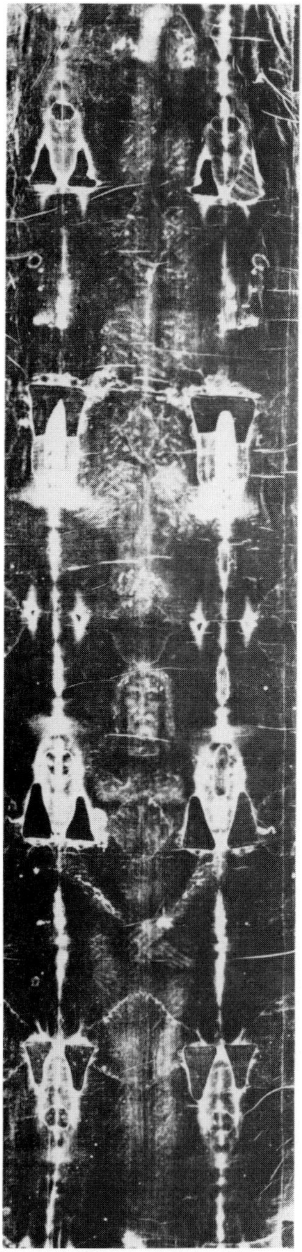

**Negative image of the Shroud**

# The Silent Witness

**THE AMAZING STORY OF THE SHROUD OF TURIN**

**The Silent Witness reveals new evidence that has startled the world.**
**Winner of 10 International film awards, this fascinating detective story examines the authenticity of the piece of cloth believed to be the material that was wrapped around the body of Jesus Christ, 2000 years ago.**
**How has it survived this length of time and why has it's compelling image bewildered so many people for centuries?**
**Forensic experts, historians, criminologists, theologians and NASA scientists confront the question**

**WHO IS THE MAN ON THE SHROUD?**

*"The great Detective thriller"*
**Daily Express**

*"Few will fail to find it riveting"*
**The Times**

*"An Amazing Film"*
*"One of the best detective stories I have ever seen"*
*"I was riveted to my seat"*
**Peter Farley, Capitol Radio**

*"Beautifully made"*
*"Technically brilliant"*
**B.B.C.**

# FILMS...

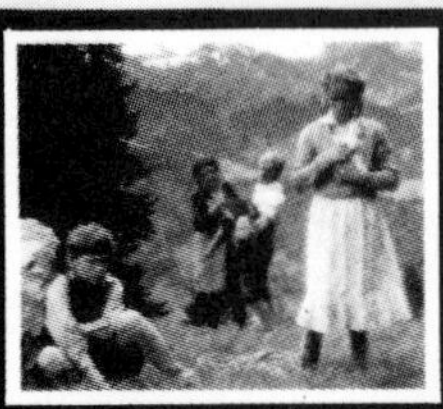

**Treasures Of The Snow**
A family film of great beauty and charm.

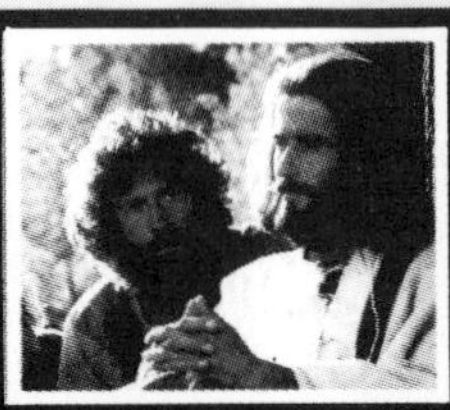

**Jesus**
A faithful recreation of Luke's Gospel.

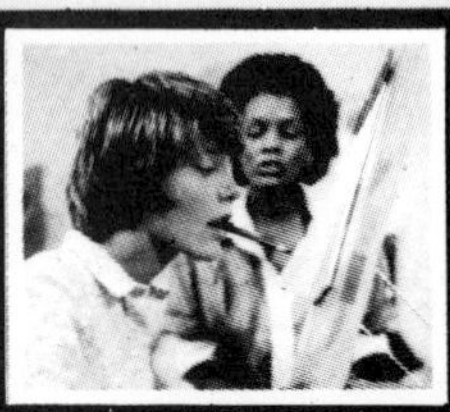

**Joni**
Joni Eareckson's gripping story of almost total paralysis.

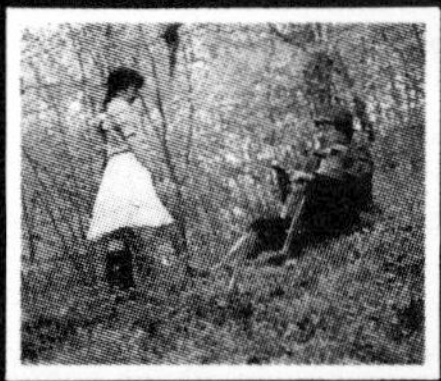

**Tanglewoods Secret**
A good story for both adults and children, with a clear Christian message.

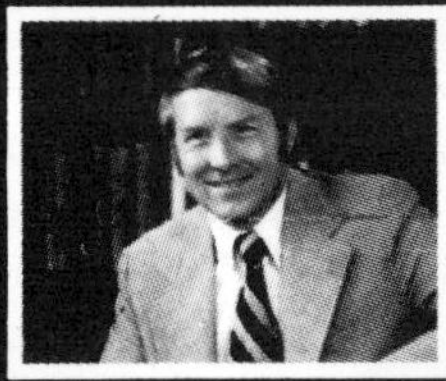

**Focus On The Family**
Dr. James Dobson's popular and greatly used series on marriage and children.

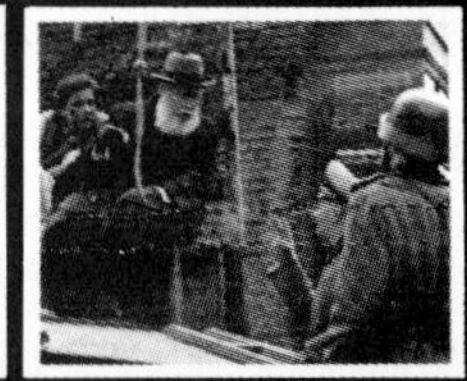

**The Hiding Place**
A true World War II story about the persecution of the Jews showing how love can overcome hate.

. . . to make you laugh, ***FILMS*** to make you think, ***FILMS*** that make you want to sing, ***FILMS*** that make you want to pray.

INTERNATIONAL FILMS have a wide variety of films for hire and some on video cassette. Films that are ideal for outreach, teaching and Church growth. 9 minute cartoons to 120 minute full length feature films PLUS teaching series such as James Dobson's 'Focus On The Family', Joyce Landorf's 'His Stubborn Love' and the'Marriage Enrichment'series.

Send £2 for illustrated catalogue to:

**INTERNATIONAL FILMS**
235 Shaftesbury Avenue, London WC2H 8EL Tel: 01-836 2255